# Smart Architecture – A Sustainable Approach for Transparent Building Components Design

Valentina Frighi

# Smart Architecture – A Sustainable Approach for Transparent Building Components Design

Springer

Valentina Frighi
Department of Architecture
University of Ferrara
Ferrara, Italy

ISBN 978-3-030-77605-3 ISBN 978-3-030-77606-0 (eBook)
https://doi.org/10.1007/978-3-030-77606-0



This Springer imprint is published by the registered company Springer Nature Switzerland AG
The registered company address is: Gewerbestrasse 11, 6330 Cham, Switzerland

# Preface

The book explores, within the world of smart architecture, the specific role that glazing technologies play in this specific framework, developing a tool to compile and quantify options for decision-making support toward the sustainable integration of them within complex building envelope systems.

Before entering the main topic, that is, indeed, that of transparent building components, the book explores the concept of the smart building envelope, as the latest goal of contemporary architectural and technological research. After defining the concept of "smart" in architecture, understanding how innovative technologies and materials have shaped buildings over years, offering technological answers to current critical needs, the work provides a supporting database of contemporary smart architecture, mapping adopted strategies, recognizing common patterns, and evaluating current and future trends in the context of the building envelope to help further development of high-potential innovative building components. Now recent breakthroughs in building technologies seem to be able to open up a range of possibilities for architects and professionals to influence design, especially as far as glazing is concerned.

Since transparent building components are still the weakest point of building envelopes, especially in terms of thermal resistance, fenestrations' design optimization challenge is more complex than for many other building components and systems. For these reasons, there is the need to provide designers with the appropriate tools to manage such complexity, according to a sustainable and integrated approach that considers the building system as a whole.

Indeed, the tool developed is aimed at encouraging the arising of relevant questions (*what-if* scenarios), approaching design problems, and suggesting a series of solutions to reduce critical issues that concern the design of transparent building components.

Therefore, the book aims to help in solving the existing lack of guidance about how the contribution given by glazing technologies could be profitably exploited within buildings, aimed at bridging the gap from current practice toward meeting future energy targets.

Further, through the book, a reader can deepen his knowledge about the state-of-the-art technologies for smart adaptive building envelopes and innovative technologies and materials thus related.

It can be expected that smart building envelopes, in general, will have a much wider implementation in the near future due to the acceptance—within the construction industry—of technological innovations of the last years. Any forthcoming directions for the development of high-performance smart building envelopes must include an improved cognition of perception elements, as well as the building industry, need to handle more detailed information beyond what is available today.

Ferrara, Italy Valentina Frighi

# About This Book

The book, after a brief recognition of how the building envelope concept has evolved over time, deals with the definition of the concept of "smart" in architecture to further examine how technology advancements change architecture, transforming architectural practice as well.

Then, the author provides a systematic characterization of recent examples of smart architecture, mapping adopted strategies, recognizing common patterns, and evaluating current and future trends in the context of smart building envelopes, energy efficiency, and the development of high-potential innovative building components. The work proceeds with a focus on the specific role that glazing technologies play in this framework, classifying existing smart windows technologies, and providing a systematic methodology to quantify options for the effective integration of transparent building components within advanced and innovative building envelope systems.

The book's structure is divided into four sections: the first deals with the reference background, giving preliminary definitions about how building envelope can be defined nowadays (according to new emerging needs) so as, how building materials, components, and systems have evolved their performance accordingly, toward a new generation of smart materials and products able to face rising issues (Chap. 1).

Secondly, the need to define the term *smart* in the specific domain of architectural technology emerged (Chap. 2), so as, in the end, the need to understand the modifications which occur over time in architecture's features, through the role that such technological evolutions have performed, in responding to current emerging needs (Chap. 3).

The second section contains the research management tools, the materials needed to conduct the research activity on which the book is based, and the methods employed to elaborate and process them. Indeed, in Chap. 4, some examples of smart architecture are presented, described, and cataloged, according to different predetermined criteria; the specific focus is on their building envelope to understand which kind of technologies or technological innovations they adopted and how this helps in conferring them the *smart* feature. Chapter 5 instead focuses on the importance of windows within building systems, highlighted by the majority of the innovations adopted in the sample investigated which address transparent building components.

The final section elaborates data gathered so far with considerations previously made: in Chaps. 6 and 7, an up-to-date state of art about glazing technologies for Smart Architecture (the so-called Smart Windows) have been provided while Chap. 8 introduces the decision support tool(box) for helping in the early stage design of transparent building components, supporting their sustainable inclusion within building systems.

# Contents

# About the Author

**Valentina Frighi** is a registered architect. She completed her PhD with the Department of Architecture at the University of Ferrara, where she is currently involved in didactic and scientific activities on topics related to architectural design and technology. Her research interests lie in the field of architectural technology, focusing on building envelopes, (innovative) building materials and components, environmental design, and building performance. She has authored and co-authored a monograph and papers for many academic and professional journals and books, collaborates on national and international research projects, is a reviewer for several journals, and has been part of scientific committees for several conferences.

# Chapter 1
# The Building Envelope

## 1.1 Definition of Building Envelope Systems

The building envelope can be defined as the main division element between environmental factors and the needs of end users, entrusted with the task to regulate interactions between inside and outside.

It is composed of external structures that delimit a building toward the outside, and its task is to act as a selective filter to guarantee constant comfort conditions of buildings' occupants. Indeed, through the building envelope, environmental and technological parameters to be considered during the whole design and construction process must come to a balanced synthesis, able to blend different and sometimes juxtaposed needs.

As Fitch (1947) stated in his main work, the building envelope—due to the high variability of the influencing factors[1] that act on it—should be imagined not as a simple barrier but as a permeable and selective membrane able to admit, reject, and filter each of the factors that act on it.

The first conditions for the development of the modern concept of building envelope were introduced by Bahnam in 1969. In his work, he recognized the role that building envelope has in maintaining determined comfort conditions, controlling and modeling interactions and exchanges that occur among inside and outside, reacting to environmental variability, and minimizing thermal losses during the winter period while limiting temperature increasing in summer.

[1] The term *influencing factors* (Conato and Cinti 2014) defines environmental and design features that characterize the setting of a single architectural project, resulting into restrictions and conditions to which building envelope has to respond in terms of performance by means of the properties of the layers and/or skins that compose it. Influencing factors can be divided into *fixed factors*, unchangeable over time, given from the contextualization of the building envelope in a specific environment, and *variable factors*, whose characteristics and attributes changeable over time.



V. Frighi, *Smart Architecture – A Sustainable Approach for Transparent Building Components Design*, https://doi.org/10.1007/978-3-030-77606-0_1

Therefore, it is clear that building envelope has to be able to respond to certain performance requirements, established by the in-force regulations—according to different application contexts and legislation—and depending on the degree of variability of external boundary conditions as well as on specific needs of buildings' intended use.

Furthermore, acting as a complex filter, the building envelope plays a central role in regulating indoor temperatures and determining the whole system consumptions, being in this way responsible for their control and reduction. For these reasons, it must be designed as a very fine-tuned system, within which each component contributes to the definition of building's global performance.

The main functions that a building envelope has to perform can be resumed as follows:

- *Support function* aimed at providing structural support against loads and forces. Although the building's facade and structures are very often detached, the envelope has anyhow the task to support its weight as well as the weight of the other building components.
- *Exchange control function* aimed to prevent undesired fluxes between inside and outside. Even if a wide variety of climatic and thermodynamic phenomena influence the behavior of a facade, the parameters that a building envelope has to face are basically temperature, air pressure, and humidity, whose variations are caused by energy transfer phenomena as a consequence of the external dynamic conditions;
- *Aesthetic-finishing function* to make the building attractive while performing the other two functions.

To promptly practice the three abovementioned functions, an effective building envelope has to actively interact with the environmental surroundings, exploiting the site's opportunities (as well as positive influencing factors while reducing boundary's weaknesses), favoring the use of renewable energy sources, and employing sustainable materials.

To give a schematic overview about principles and parameters that have to be considered during building envelopes' design, it can be said that an efficient and effective building envelope has to:

- passively control *solar incident radiation*, interacting with it to avoid undesired overheating during summer while maximizing solar heat gains in the winter season
- enhance *daylight*, reducing the needs of artificial lights
- modulate and control *natural ventilation* to increase passive cooling in hot seasons to ensure a good level of indoor air quality
- deal with *thermal mass storage* and *energy transfer*

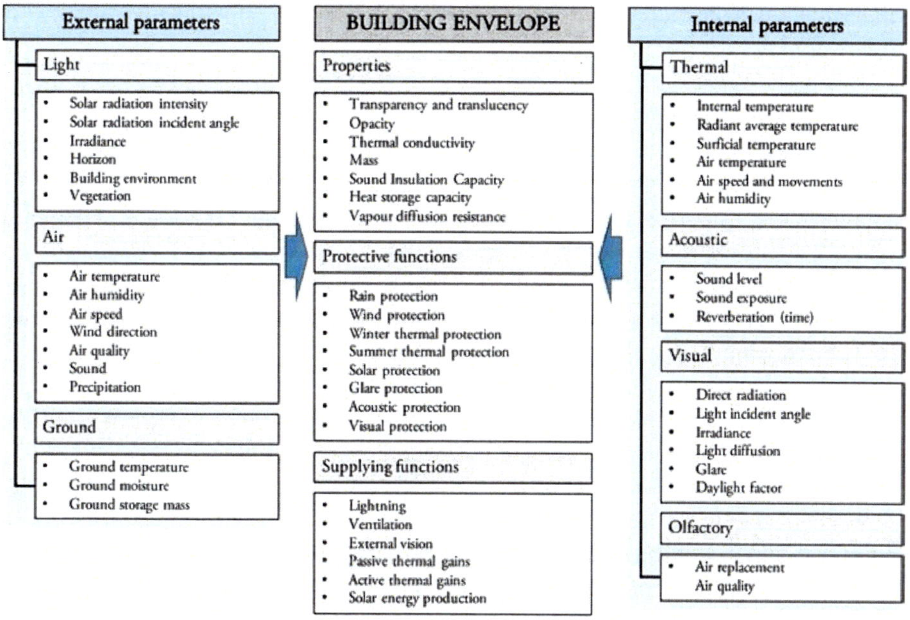


**Fig. 1.1** Synoptic framework of external and internal influencing factors (and relative parameters), features, and functions of building envelope. For further insights, refer to Frazzica (2008)

- guarantee an appropriate *thermal insulation level*, controlling heat transfer through it
- guarantee an appropriate *acoustic insulation level*
- guarantee the *hygrometric–air level modulation* and *condensation control.*

Additionally, essential requirements are *water* and *wind tightness*, *maintenance and repair facilities*, and *fire safety* (Fig. 1.1).

Therefore, it can be said that the typical design needs that building envelopes have to fulfill can be ascribed to six main families that summarize the abovementioned features, which are: (1) thermal control; (2) energy performance; (3) indoor air quality; (4) visual performance; (5) acoustic comfort; and (6) interaction and users' control.

## 1.2 The Building Concept in a Changing World

> The design of the building envelope has to address a wide spectrum of issues, ranging from the technical performance of the individual materials and the nature of their assembly to the visual appearance and propriety of the resulting building form.
>
> Bob Allies, introduction to Lovell (2010)

Building concept, and, with it, features and requirements of building envelope systems, has incredibly changed over the years. If once buildings were only less more than a shelter, now they have become the places in which we spend most of our everyday life, working, and living.

So, based on what kind of activities they host, buildings have to be promptly designed in both layouts as in performance terms.

Indeed, a building envelope can be considered efficient if it can successfully respond to all the *design needs* identified in the previous paragraph (see Sect. 1.1); this means considering that surrounding influencing factors in actual application contexts are too often randomly changing.

For this reason, this condition must be satisfied by resorting to a dynamic interface, continuously interacting with external climatic conditions and able to respond to external stimuli as well as to the ever-changing needs of buildings' end users.

To manage the complexity of multiple factors that influence buildings' behavior, a building system should make use of different and complementary contributions and continuous interactions among components that made it up.

Therefore, if once building envelopes were conceived only as *monolithic* shells, now they have to be designed as *sets of layers* (much lighter than in the past) equipped with specific tasks to mediate external climatic conditions with people's internal comfort perception through their cooperation. Clearly, it follows that the overall quality of a building is largely related to the technological features of its closures; for this reason, modern building envelopes have to be intended as *advanced systems* that must consider simultaneously architectural and urban constraints, structural performance, energy efficiency, and indoor environmental quality.

So, following the well-known energetic and environmental questions, as well as the arising of new dynamics that radically changed project's demands, it seems no longer possible to rely only on building envelopes equipped with static performance—although high in their absolute values—to adequately respond to the needs of extreme variable contexts, but instead, we should aim at developing technological solutions suitable for the increasing complexity of different application's sites, able to adjust and to adapt themselves to the continuous variations that occur. This seems to lead to the adoption of *dynamic* devices and systems equipped with greater possibilities, thanks to their adaptation capacities.

## 1.3 Requirements and Parameters

### *1.3.1 Influencing Factors*

The building envelope, acting as a filter between the internal environment and external surroundings, must be appropriately designed to respond to the application context in which it will be inserted. Indeed, on a building, and especially on its envelope (made of opaque and transparent parts), acts different external factors and agents, defined as *influencing factors*.

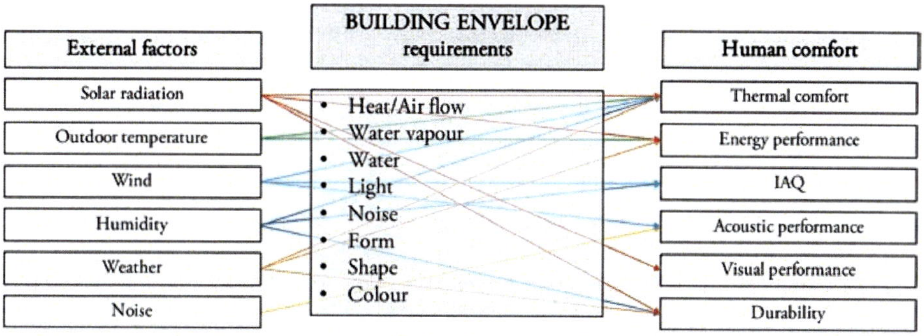


**Fig. 1.2** Building envelope requirements in the relationship between external factors and human comfort needs. For further insights, refer to Aelenei et al. (2016)

Influencing factors can be distinguished from *fixed factors* and *variable factors* and, again, from *external factors* and *internal factors*, depending on the application context to which they refer. Fixed factors are those that are unchangeable over time—coming from the specific geographic context—while variable factors are those with transitionary features.

As one can imagine, all the influencing factors affect human comfort in multiple ways; the factors on which comfort depends mostly are basically the function of the geographic context, such as climatic area, orientation, site's morphology, outdoor temperature and humidity, wind, and precipitations as well as the relationship with other buildings (able to determine different amounts of solar radiation), to name a few (Fig. 1.2).

By virtue of the fact that each building site is necessarily exposed to different factors, every facade should have a customized typological, technological, and performance arrangement to optimize potential interactions among climatic features and energetic issues, although it is clear that it could not be a feasible alternative in some applications. For this reason, the will to include the dynamic dimension of nature in architectural technology has emerged, leading building envelopes to establish continuous interactions with constantly changing boundary conditions.

### *1.3.2 Building Envelope Requirements*

The abovementioned influencing factors to which a building envelope should appropriately respond can be translated into *parameters* and *requirements* to be fulfilled to ensure the internal environmental comfort of a building and, consequently, its correspondence to current regulations and needs of end users.

The *internal environmental comfort* of a building is conventionally defined as the condition in which a human being is satisfied toward the environment that surrounds him. It expresses a neutrality condition and is completely personal, being largely influenced by physiological subjective factors (such as metabolism, ages, clothes,

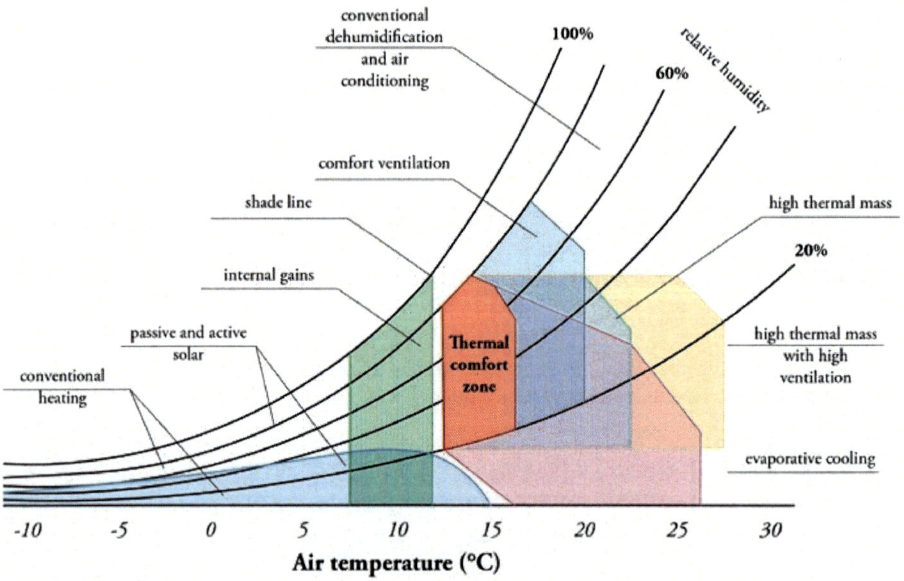


**Fig. 1.3** Psychrometric chart that shows the effectiveness of different design strategies in extending the potential comfort zone. For further insights, refer to Lovell (2010)

physical activities, etc.) that are not quantifiable and by other factors (such as temperature, humidity, airspeed, etc.) that, on the contrary, are countable environmental parameters.

So, the external environmental perception is obtained through the processing of a set of thermal, visual, acoustic, olfactory, and psychological inputs, filtered and altered due to an individual sensitivity which is, in turn, influenced by heat exchanges (for conduction, convection, and radiation), the body metabolism, and the evaporative processes determined by breath and skin perspiration (Fig. 1.3).

In general terms, the variables that affect the buildings' thermohygrometric comfort are:

- the *air temperature*, usually measured in °C
- the internal air *relative humidity*, expressed in percentage, which is the expression of the ratio of the vapors' quantity in a mass of air and the maximum amount that such mass can contain at the same temperature and pressure conditions
- the *average radiant temperature*, generally expressed in °C and calculated as the mean value of internal walls' temperature, ceiling and floor included
- the *airspeed*, expressed in m/s.

Although, as previously said, physical parameters that influence the environmental comfort are subject to a personal perception, minimum requirements' values that indicate general and commonly shared accepted comfort conditions are stated (generally on a statistic base) in national or international norms UNI EN ISO Norm 7730:2006 (Figs. 1.4 and 1.5; Table 1.1).

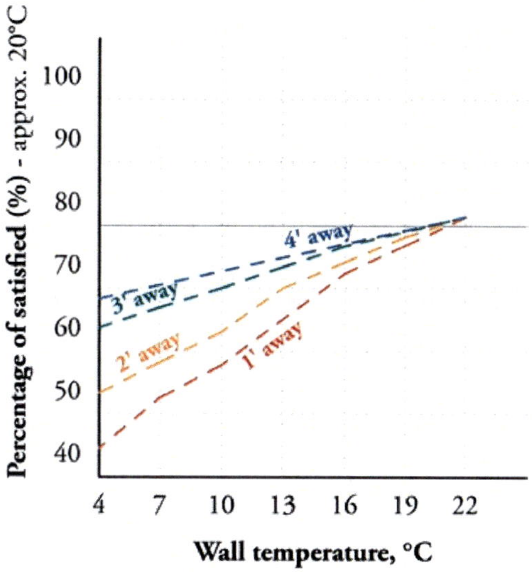


**Fig. 1.4** Trend of human comfort (in terms of percentage of satisfaction) in relation to the proximity to a cool surface. Author's elaboration on Building Science Consortium Data

In general, the requirements that building envelope has to fulfill can be briefly divided into two categories:

- *Mandatory requirements*—those defined by local, regional, and national regulations we are obliged to respect to ensure a minimum level of performance
- *Discretionary requirements*—that can be reached exceeding the previous ones with advanced design options or construction practices, promoted to improve the performance of the building system as much as possible.

Among such groups, we can do one more subdivision by distinguishing requirements based on the main field to which they pertain. So, we can recognize *technological requirements* intended as those that comprise all the issues whose control is entrusted to the building envelope and whose minimum values to be respected are generally defined by regulations. They concern energetic, visual, and acoustic issues and, more in general, all the requirements of technological nature, even if specifically directed toward plants or other specific parts of the system.

On the other hand, we can recognize *typological requirements* that have to be fulfilled mainly through the layout of the rooms in relation to the external environment (orientation, wind, sun, etc.). They basically concern user-related issues, both in terms of layout and enjoyment of the spaces and in control and management terms.

So, to fulfill both the abovementioned requirements, a well-designed, efficient, and effective building envelope has to:

- ensure the comfort within confined spaces during juxtaposed seasons
- guarantee a correct thermal and acoustic insulation level as well as the correct installation of its components to avoid thermal bridges that could compromise its performance

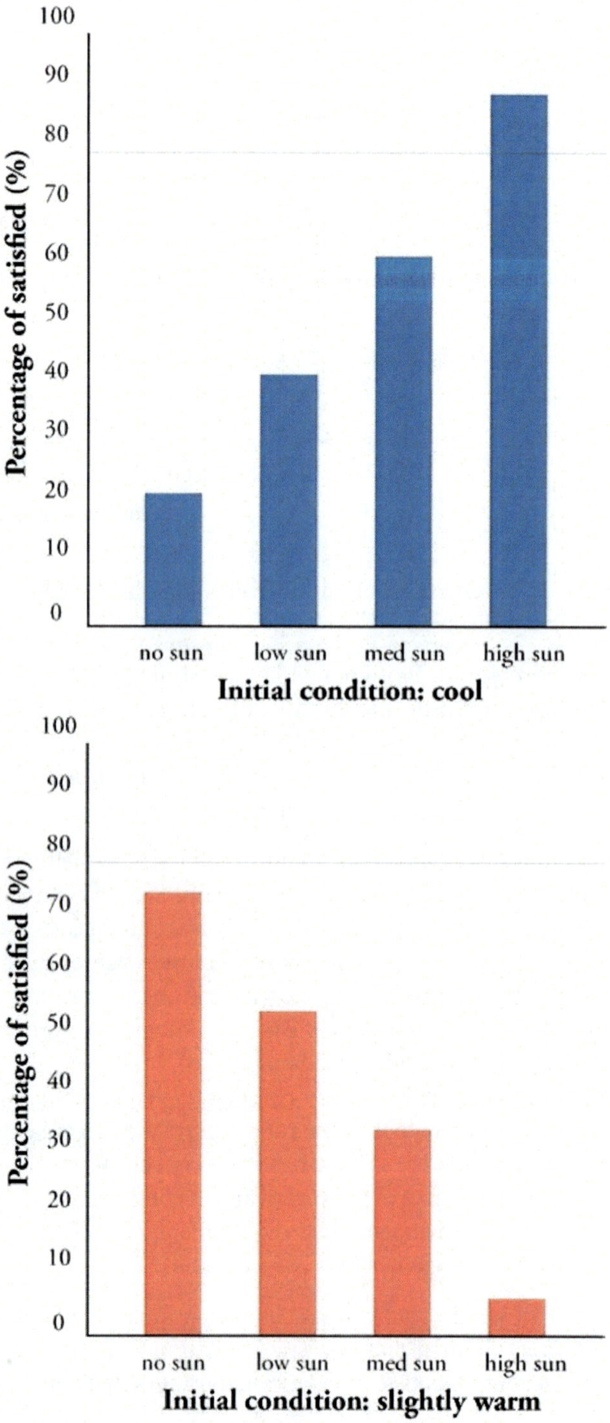


**Fig. 1.5** Relationship between human comfort (in terms of percentage of satisfaction) and solar gain in winter (*above*) and summer seasons (*below*). Author's elaboration on Building Science Consortium Data

**Table 1.1** Seasonal comfort conditions coming from the recommendations included in the UNI EN ISO 7730 for a person in a sedentary state or under light physical activities

| Variable | Winter season | Summer season |
|---|---|---|
| Operating temperature | 20–24 °C | 23–26 °C |
| Relative humidity | 30–70% | 30–70% |
| Airspeed | 0.15 m/s (22 °C) | 0.17 m/s (24 °C) |

- provide an appropriate number of openings, properly treated and shaded to maximize positive solar heat gains during cold seasons and to avoid overheating during hot seasons
- adopt effective ventilation strategies
- make use—as far as possible—of renewable sources for energy production purposes and
- exploit building automation strategies to optimize operation modes and functioning of the whole building system, even according to the habits of end users.

Last but not least, it has to consider the continuously evolving relationship among architecture, humans, and environment.

To give an overview of what generally are the major requirements that building envelopes have to satisfy, as well as relative reference parameters that as to be controlled during the whole process, primary technological and typological building requirements are here briefly described.

## Technological Requirements

As stated earlier, technological requirements attain mainly comfort perception[2] within buildings, both from a thermal and from an acoustic point of view. They are strictly related to material properties, form, and fabrication that compose the building envelope and can be grouped under the energetic and acoustic definition.

### Energetic Issues

Energetic issues comprise parameters and requirements that mainly pertain to thermal and thermo-hygrometric control, which has to be guaranteed by building components.

Particularly for transparent component, energetic issues are significant and mostly influenced by incident solar radiation that determines phenomena able to increase undesired thermal loads within rooms (such as, the greenhouse effect) and consequently serious challenges to deal with (Fig. 1.6).

[2]Human comfort sensation within confined spaces can be generally considered verified when the internal temperature is comprised between 24° and 25° during summer and among 18° and 22° during winter. Clearly, this perception depends on a subjective sensation, influenced by variables such as the metabolism, the clothes level and the degree of physical activities.

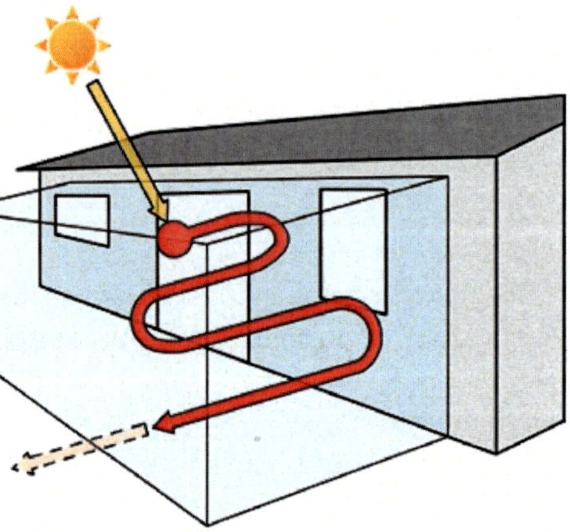

**Fig. 1.6** The greenhouse effect: Part of the solar energy (UV + light + short-wave IR) entering a room through a transparent component is absorbed by air and objects surrounding, raising their temperature. The objects, in turn, give off heat (long-wave IR) that does not filter out through the glass. Due to this accumulation of heat, the air temperature will rise significantly.

The ability of the materials and components to retain heat and to tackle temperature peaks—in other words, their *thermo-energetic behavior*—is expressed through physical parameters, such as thermal transmittance, thermal resistance, heat conductivity,[3] heat capacity[4], and specific heat, thermal lag, and frontal mass for opaque layers.

On the other hand, as anticipated before, the definition and description of the energy performance for transparent components are quite more complex due to their possible interaction with incident solar radiation; for this reason, they have to be described in terms of both thermal and optical parameters (Fig. 1.7).

The energy performance level of a building envelope is currently evaluated based on its correspondence to requirements established within national, regional, or local regulations, which, in turn, for the European Community, descend from common directives in terms of energy efficiency.

Until now, the key laws that cover the reduction in buildings' energy consumption, resulting in parameters and requirements that have to be respected, are the *2010* Energy Performance Buildings Directive (*EPBD*) (European Commission, 2012) and the *EED 2012* (otherwise known as EPBD Recast), amended in 2018 (European Commission, 2018) as part of the "Clean energy for all European" package (European Commission, 2019) to update the policy framework to 2030 and

[3]The heat conductivity, or thermal conductivity (λ [W/m°K]), represents the insulation performance of a material, expressed through its ability in transmitting heat. The lower is the value of the heat conductivity, the better will be the heat insulation performance of the material.

[4]The heat capacity [J/K] of a material expresses its ability in storage heat to further release it to the surrounding environment. This feature is particularly important during summer season thanks to the attenuation and delay of heat entering in rooms; on the other hand, during winter, this ability allows to diminish building's internal thermal fluctuation.

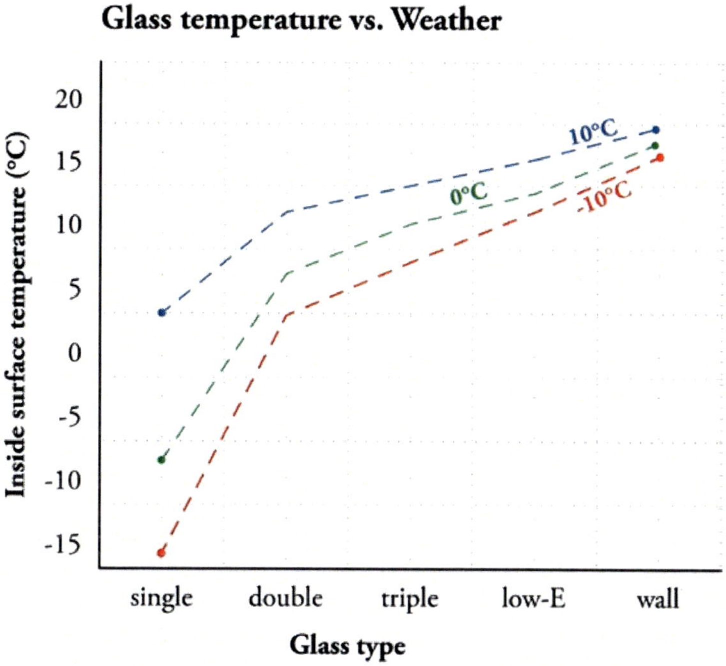


**Fig. 1.7** Relationship between glass type and outdoor temperature trend with internal surface temperature variations. Author's elaboration on Building Science Consortium Data

beyond. Although significant is the update proposed by the Commission in 2016 to promote the use of smart technology in buildings[5] and the agreement reached on such update at the end of 2017 about the addition of a series of measures to the current Directive "aimed at accelerating the cost-effective renovation of existing buildings. There will also be updates to provisions on smart technologies and technical building systems" […].

For opaque envelopes, we can say that, in general, the energy requirements that have to be fulfilled—according to the previously mentioned regulations—are:

- the *thermal transmittance*, or U-value, expressed in W/m$^2$K;
- the *frontal mass* intended as the mass for the surface's unit of the opaque wall, expressed in kg/m$^2$. This parameter has been introduced to avoid energetic evaluations only under stationary conditions, trying to simulate the energy behavior of a wall even in dynamic conditions, thus limiting the effects of external temperature variations within confined spaces;

[5] On the 30th of November 2016, the Commission proposed an update to the Energy Performance of Buildings Directive to help and promote the use of smart technology in buildings and to streamline the existing rules. The Commission also published a new buildings database—the EU Building Stock Observatory—to track the energy performance of buildings across Europe. More information and materials about are available at the following link: https://ec.europa.eu/energy/en/news/commission-proposes-new-rules-consumer-centred-clean-energy-transition

- the *thermal periodical transmittance* ($Y_{IE}$) expressed in $W/m^2K$ and intended as the ability of an opaque element in displacing and attenuating the thermal flux that passes through it in a day. It basically quantifies the inertial capacity of a component, which means its ability in diminishing internal temperature variations when compared to external temperature during the summer season as well as to store heat during the day to further release it gradually during the night. This value is strictly connected to the frontal mass of the component.

Another parameter that influences quite a lot the energetic behavior of a building is the amount of natural light within rooms, coupled with its correct balance with the artificial compound. If a building presents the correct morphology, dimension, and position of its opening, we can say that, in general terms, it will be able to maximize the natural lighting supply.

However, to ensure lighting comfort, the natural compound has to be accurately designed according to the use of spaces, exploiting even the reflected and diffused light components and resorting to particular devices, if necessary.

Such requirements are exclusively ensured by the presence of transparent building components for which, generally, different parameters are established and aimed at ensuring a correct level of natural lighting within confined spaces. One of them is the *average daylight factor*, which establishes minimum standards according to the destination of use. For residential rooms, the average daylight factor is generally fixed around 2%; it is expressed in percentage as the ratio among the average illuminance of a confined space, given by natural light, and the external illuminance on a horizontal surface, exposed to the vaulted sky, without any obstruction, in an overcast condition.

Other parameters involved in the definition of the environmental comfort related to transparent building components are those connected to the amount of light passing through the glazing, thus indicating the selective/reflective ability of the glass employed. Unfortunately, these values (transmission coefficient, solar heat gain coefficient, or light to solar gain ratio) are too often a prerogative of designer and technicians' best practices because they are not always taken into consideration in regulations.

### Acoustic Issues

Speaking about technological requirements cannot avoid mentioning acoustic issues. Indeed, the acoustic well-being of a room is a fundamental parameter to guarantee the proper development of activities within it.

The acoustic quality of an environment depends not only on the presence or absence a disturbing sound source but also on many other factors, the function of the environmental features of the surrounding, the level of eventual background noises, and the constructive and technological building's features.

So, if on one hand correct design and installation of building components have to be properly done to control such issues, on the other hand, even the correct arrangement of rooms within the building—that must occur after the evaluation of the presence of actual or potential sound sources, predisposing, if needed, protected views for sensitive uses, and/or exploiting distribution spaces as buffer spaces—can help for the achievement of an adequate comfort level.

The evaluation of acoustic comfort normally happens by referring to the sound level, that is, a maximum threshold of acceptable noise that does not disturb a person within the environment.

Sounds, indeed, spread in air or through materials with different speed, according to two different mechanisms:

- via *air transmission*, where sound spreads freely in the air without facing solid barriers
- via *structural transmission*, where sound spreads through the solid structure through elastic vibrations.

In addition, in building components, it is necessary to distinguish between direct transmission, which happens when sound energy transmission occurs only through the considered components, and flanking transmission, which happens when the transmission occurs through structures near the one considered.

Acoustic insulation is generally defined as the difference between the average sound pressure values between adjacent rooms: one room in which is located the sound source and the receiving room in which is located the receiver; such value is not an intrinsic property of building components and instead it depends on:

- the acoustic properties of single components
- the mechanical properties of bearing structures
- the acoustic absorbing features of internal surface materials that determine the sound reflection within the two environments.

### Typological Requirements

As mentioned previously, the environmental comfort within confined spaces is strongly related to rooms' layout, in relation to both the functions that take place in them and the climatic and micro-climatic external conditions. Therefore, arrangement of spaces must take into account the needs related to thermal, lighting, and acoustic comfort as well as toward activities carried out within them.

Even building orientation must be carefully considered to maximize solar positive contributions where necessary while passively acting on heat losses, arranging buffer spaces toward disadvantageous views.

For this reason, diversified thermal regimes that occur at various orientations, due to different daily irradiation conditions, should be reflected by the internal distribution of environmental units.

To generally resume common best practices, we can say that on north orientation, temporary environmental units and buffer spaces that can protect main rooms from external thermal variations can be located; on south orientation, long-term main environmental units that require high-thermal and light comfort should be appropriated; and on intermediate orientations (east/west), long-term secondary environmental units, used for a particular time of the day, would be fine.

All the abovementioned questions must be considered even in relation to the new ways of life, habits, and family settlement of end users, far from being static over time due to the constant transience and fluidity of such VUCA[6] reality.

## References

Conato F, Cinti S (2014), Architettura e Involucro - II Edizione. Be-Ma Editrice, Milan

European Commission (2012), Directive 2012/27/EU of the European Parliament and of the Council of 25 October 2012 on energy efficiency, amending Directives 2009/125/EC and 2010/30/EU and repealing Directives 2004/8/EC and 2006/32/EC Text with EEA relevance. https://eur-lex.europa.eu/legal-content/EN/TXT/?qid=1399375464230&uri=CELEX:32012L0027 Accessed 10 Feb 2021

European Commission (2018), Directive (EU) 2018/2002 of the European Parliament and of the Council of 11 December 2018 amending Directive 2012/27/EU on energy efficiency (Text with EEA relevance.). https://eur-lex.europa.eu/legal-content/EN/TXT/?uri=uriserv%3AOJ.L_.2018.328.01.0210.01.ENG Accessed 10 Feb 2021

European Commission (2019), Clean energy for all Europeans. https://op.europa.eu/o/opportal-service/download-handler?identifier=b4e46873-7528-11e9-9f05-01aa75ed71a1&format=pdf&language=en&productionSystem=cellar&part= Accessed 10 Feb 2021

Fitch J M (1947), American Building: The Environmental Forces that shape it. Oxford University Press

Frazzica G (2008), Atlante della Sostenibilità. UTET, Milan

Lovell J (2010), Building Envelopes. An Integrated Approach. Architectural Briefs, Princeton Architectural Press, New York

UNI EN ISO Norm 7730:2006 - Ergonomics of the Thermal Environment - Analytical determination and interpretation of thermal comfort using calculation of the PMV and PPD indices and local thermal comfort criteria

[6]VUCA is an acronym used to describe or reflect on the volatility, uncertainty, complexity and ambiguity of general conditions and situations.

# Chapter 2
# The Smart Concept Definition

## 2.1 Definition of the Smart Concept in Architecture

Even if the idea of "smart" is far from being newly born, since it takes, in its basic form, inspiration from nature (where all living organisms possess stimulus-response abilities), the application of this concept to the architectural field is quite recent, having appeared in the literature only at the end of the last century to define materials and components with unconventional features or selective and specialized performances opposite to traditional ones.

However, due to the multitude of existing definitions about what *smart* senses,[1] coupled with the fact that, especially in Architecture, this concept appears still fluid; Addington and Schodek, formerly in 2005, recognized that it can be employed without a precise definition of its meaning because it appears surprisingly difficult.

Today, the smart feature in the architecture's domain is generally associated with materials and technologies highly engineered, able to respond in an intelligent way to the context in which they are inserted if subjected to external stimuli, as well as to vary their properties, change their structure and/or form, or to assume different functions (Drossel et al. 2015).

Thus, referring to smart materials and technologies means speaking about something with an advanced and sometimes dynamic response, by its nature juxtaposed to "traditional" materials and technologies, generally equipped with mainly static performances.

Nevertheless, it has to be said that there is a very fine line that divides "smart" materials from "conventional" materials (Bogue 2012, 2014) because a lot of existing materials can react or respond to some stimulation (Oliveira et al. 2018) to

[1] See, for instance, the definitions provided within Addington and Schodek (2005) or, again, the others in Kroschwitz (1992) and Drossel et al. (2015).



V. Frighi, *Smart Architecture – A Sustainable Approach for Transparent Building Components Design*, https://doi.org/10.1007/978-3-030-77606-0_2

behave in a smart manner; this is even more true in Architecture, where the global performance of the buildings largely depends on the overall behavior of a set of heterogeneous components.

So, in the specific field of architectural technology, different interpretations of such concept seem possible, which go from the single building material to the component—in which heterogeneous materials and products relate each other in a "smart" way—up to the building envelope system in its whole, in which different components work together to provide a dynamic and tailor-made performance response, according to the variability of boundary conditions (Fig. 2.1).

Therefore, the *smart* concept has not to be intended only as the integration of different or unconventional features in a given material, component, or system, but it can also be attributed to the intelligent application of something within the building system in its whole or, again, to the intelligent management of a complex organism such as the building.

So, it is clear that, especially in the specific domain under investigation, several complementary definitions of the concept of *smart* seem possible and suitable for the different application scales that mark out the building process.

Accordingly, such an idea led to the definition of *smart buildings* as systems able to adapt, in a responsive way, not only to the variability of the boundary conditions but also to the continuously evolving needs of their end-users.

Casini (2016) indeed defines a *smart building* as the evolution of a nearly zero energy building in which to the high and static performance of employed materials and technologies, the possibility of dynamically controlling the building's environmental parameters, depending on the variation in external or internal conditions, is added.

**Fig. 2.1** Sawmill, Tehachapi, California (USA). Kevin Scott/Olson Kundig. A family retreat designed as a low-tech smart home, meaningful for the smart definition here provided. In Sawmill, a net-zero home, the living room's facade can transform and retract without recurring to advanced technologies but just employing a complex system of cranks and levers. (Courtesy of Olson Kundig)

So, a smart building is not only an edifice that makes use of building automation technologies—definition which sounds as rather reductive—but it is a building equipped with technological solutions able to provide high performance in terms of comfort, energy efficiency, and environmental sustainability, to dynamically interact with the surrounding, and to acquire data and information useful for optimizing its functioning, thus correcting possible inefficiencies.

Despite such quite common agreement, literature on smart buildings is somewhat lacking, especially regarding the formulation of a precise shared definition, as each author emphasizes certain issues over others.

A definition that seems suitable for the present work is based on the performance that architecture must ensure to appropriately respond to actual standards through its envelope, which basically can be resumed as follow:

- heat-insulation control
- acoustic control
- sun protection, during hot seasons
- daylight control and active shading, when needed
- maximization of SHG, during cold seasons
- active air and ventilation control
- energy production, through renewable sources
- indoor air quality control
- passive design strategies and exploitation of site's resources
- reports and data on real performance, during operation time
- self-adjustment capabilities based on users' feedbacks or predetermined input sceneries.

Evidently, this entails the adoption of new, efficient building materials and components as well as of information and communication technologies (ICT) to perform monitoring and controlling functions (OECD 2009), such as, for instance, temperature sensors and heat detectors, light-level detectors, movement and occupancy sensor, smoke and gas detectors, and status sensors (e.g., IAQ, open windows, breaking sensors, etc.).

## 2.2 Applying ICT in Smart Buildings

The rapid development in ICT over the last decade created the technical basis for the current implementation of Industry 4.0; this mainly refers to the transformation from *automated* to *intelligent* manufacturing, which determined the networks' extension to real objects and things,[2] allowing new unconventional and interesting possibilities even in architecture.

Indeed, the applications of building sensors and wired objects, as well as their coupling with smarter control systems, are forming the basis for smart buildings that generally make use of such technologies to optimize construction, monitoring, and management costs.[3] Examples of smart controls are promising and reporting significant savings: from 10% for simple controls up to 60% for the most advanced (Lindera et al. 2017) (Table 2.1).

Since recently, the management of building systems has become ever complex, and the domotic market has increasingly spread, especially in the residential sector, where ICTs are constantly growing.

Two projections, among others, expected that by 2026, the smart building market will reach USD 229.10 billion (Mordor Intelligence 2020) and that by 2030 every major office building in a city will be powered by a smart building platform (Simonetti 2017), lowering costs and efforts by taking advantages of powerful cloud-based data analytics.

This will probably progressively result in estimated savings of up to 150 Mtoe (1745 TWh) per year by 2028 (Casini 2016), which means about 22% of all buildings' energy consumption and 9% of total final energy consumption of the entire European Union (Figs. 2.2 and 2.3).

The value of IoT is not only in enabling technologies but rather in the benefit that collected data can provide to make processes more efficient, reducing costs and managing things and people in a complex environment.

**Table 2.1** Smart Buildings' automation technologies. Author's editing based on OECD (2009), Smart Sensor Networks: Technologies and Applications for Green Growth. Available at: DSTI/ICCP/IE(2009)4/FINAL

| | HVAC | Lighting | Shading | IAQ | Switching devices | Security and safety |
|---|---|---|---|---|---|---|
| Temperature and heat detectors | ✓ | | ✓ | | | ✓ |
| Light level detectors | | ✓ | ✓ | | ✓ | |
| Movement/occupancy sensors | ✓ | ✓ | | ✓ | ✓ | ✓ |
| Smoke and gas detectors | ✓ | | | ✓ | | ✓ |
| Status sensors | ✓ | ✓ | ✓ | ✓ | ✓ | ✓ |
| Glass break sensors | | | | | | ✓ |

[2]The so-called Internet of Things, that is—in other words—the connection of almost all types of devices into a network through the web, often without the interference of man. IoT should be not treated as an independent technology but as a combination of many technologies and aspects that complement each other (Magruka 2015).

[3]Talon and Strother (2017), among others, estimate that the adoption of Building Automation Technologies or Building Automation Control Systems would allow to obtain within 2028 a saving of more than 150 MToe (the equivalent of 17.455 × 108 kWh) per year.

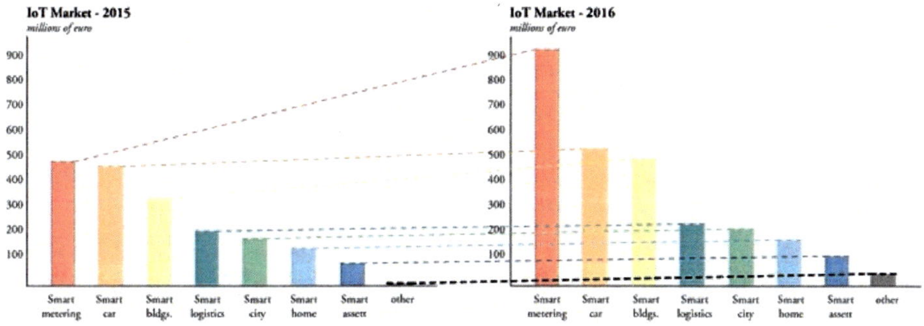


**Fig. 2.2** IoT Market's growth forecasts. Author's editing based on Internet of Things observatory data, School of Management, Politecnico di Milano

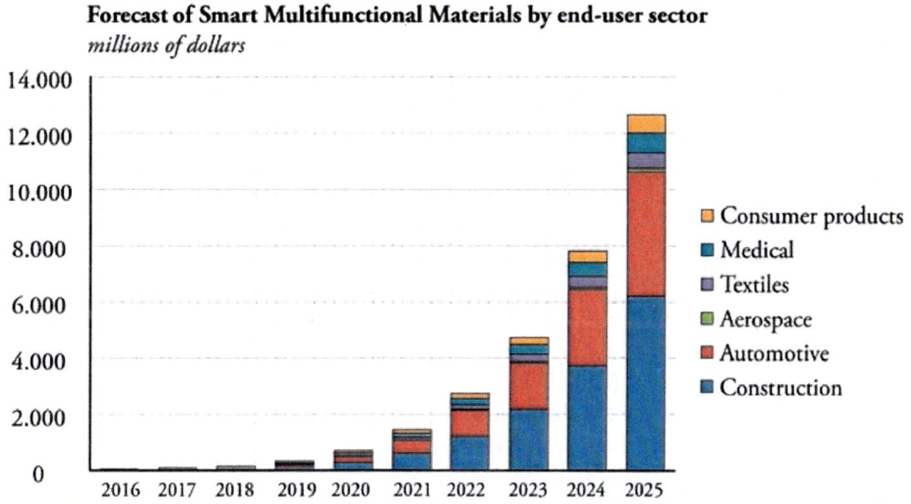


**Fig. 2.3** Forecast of Smart Multifunctional Materials. Author's editing on n-tech Research data (2016)

Most of the actual widespread technologies are currently based on advanced ICT systems and make use of shared integrate data models, such as the abovementioned cloud technologies (Fig. 2.4). With the general "cloud" term, it is intended as a technology that allows organizations to outsource data storage, communication, and/or processing (Leblond and Shoumi 2016) which comprises consumer and business products, services, and solutions delivered and consumed in real time over a network.

Cloud-based solutions are easy to deploy, able to collect and process data from anywhere at any time, deliverable to any user platform, and easily scalable and expandable.

As a result, clouds are ideal platforms for building management and automation because they can gather a great amount of data about HVAC, lighting, and energy, collected from different buildings and at regular intervals, which can be further employed to implement the same systems in new and improved applications of them.

| Cloud terminology | Definition |
|---|---|
| Public Cloud | "Multi-tenant environment" where the computer resources used are shared with other tenants. A Public Cloud is cheaper than a Private Cloud but gives the user no control over the actual systems. |
| Private Cloud | Defined as "Single-tenant environment" in which each client has systems and hardware dedicated to their applications. This costs more than a Public Cloud but provides greater control over technologies uses and ownership of security. |
| Hybrid Cloud | Environment that combines both Public and Private Cloud resources. In a Hybrid cloud, the Private Cloud is reserved for sensitive data and applications while the Public Cloud is used where possible to reduce costs. |

**Fig. 2.4** Definition of Cloud terminology. Author's editing on Leblond and Shoumi (2016)

Unfortunately, clouds that group building data about real consumption are still not so widespread because of the complexity of collecting and gathering such a variety of information. Having a meaningful sample, indeed, means to collect data about so many different parameters (representative of the specific features investigated) for an enormous period, which implies a huge quantity of information to be collected, gathered, and systematized to be further reused for any kind of activities.

The datasets of existing buildings are no more than a set of 10; they assemble information that goes from building energy consumptions to building occupations, temperature, relative humidity, and so on. Furthermore, datasets generally involve commercial and business buildings while residential ones are much less frequent; the reason why data about the residential stock are scarce depends on the complexity to acquire them, due to the heterogeneous of properties, and on the scarce availability of users in making their home accessible to install measuring devices.

Moreover, all the abovementioned information, to have significant and scientifically reliable results, would be put in relationship with estimated and real consumption data not always available.

The current most significant datasets are certainly those promoted and developed by the U.S. Department of Energy, such as the BPD (Building Performance Database),[4] the nation's largest dataset of information about the energy-related features of commercial and residential buildings. This database combines a lot of data collected by different institutions and bodies to make them available to the public.

Another significant attempt is the one made within the European Project ODYSSEE-MURE,[5] coordinated by the French agency ADEME[6] and supported by the Intelligent Energy Europe program, which involves the national energetic and environmental authorities of the 28 European countries as well as Norway, Switzerland, and Serbia and to which has taken part even ENEA, the Italian National

[4] Project website: https://bpd.lbl.gov/ and https://bpd-info.lbl.gov/

[5] http://www.odyssee-mure.eu (The access to the database is free for all EU Ministries, Concerted Action EED, EED Committee Members and EU Universities as well as research centers for non-commercial uses and via subscription for other users).

[6] Agence De l'Environnement et de la Maîtrise de l'Energie (Agency for the environment and the energetic management)

Agency for the New Technologies, Energy, and Sustainable Economic Development. It is a dataset adopted by the European Commission (EC) as a statistic fundamental for energy policies that collect energy consumption and energy efficiency as well as $CO_2$ indicators using two complementary databases: ODYSSEE and MURE, which contain policies and measures of energy efficiency and their impact since the 1980s. Meaningful of the abovementioned complexity is the sample disclosed by Italy: 85 provisions of which 65 are actually in force, a clear expression of the great amount of data to be collected, and, at the same time, to be respected over time.

In addition to the ODYSSEE-MURE database, six data tools have been developed to monitor market-penetration signs of progress of a selection of energy-efficient technologies and practices to identify drivers responsible for the variation in energy consumption to further compare and benchmark the energy efficiency and performances of countries.

In this direction are even going research centers and units, urged by the Commission; proof of what claimed is the EU-funded project ExcEED,[7] coordinated in Italy and aimed at "establish a robust and durable return of knowledge mechanism, collecting actual buildings energy performance data, and providing information to designers, energy managers and policymakers".

## References

Addington M, Schodek D (2005), Smart Materials and New Technologies. For architecture and design professions. Architectural Press, Elsevier, Oxford (UK)

Bogue R (2012), Smart materials: a review of recent developments. Assembly and Automation 32:3–7

Bogue R (2014), Smart materials: a review of capabilities and applications. Assembly and Automation 34:16–22

Casini M (2016), Smart Buildings. Advanced Materials and Nanotechnology to Improve Energy-Efficiency and Environmental Performance. Woodhead Publishing Series in Civil and Structural Engineering: Number 69, Elsevier, Oxford (UK)

Drossel W G et al (2015), Smart3 – Smart Materials for Smart Applications. Paper presented at the CIRP - 25th Design Conference for Innovative Product Creation, Haifa, Israel, 2-4 March 2015

Kroschwitz J (ed) (1992) Encyclopedia of Chemical Technology. John Wiley & Sons, New York

Leblond S, Shoumi S (2016), Leveraging Cloud-based Building Management Systems for Multi-site Facilities. Schneider Electric – White Paper 998-2095-11-05-15AR0, Rev 0. Available via https://download.schneider-electric.com/files?p_Doc_Ref=998-2095-11-05-15AR0_EN. Accessed 27 Oct 2020

Lindera L et al (2017), Big Building Data - a Big Data Platform for Smart Buildings. In Energy Procedia 122:89-594

Magruka A (2015), The most important aspects of uncertainty in the Internet of Things field – context of smart buildings, in Procedia Engineering 122:220 – 227

Mordor Intelligence (2020) Smart Building Market - Growth, trends, Covid-19 impact, and forecasts (2021 - 2026) report. https://www.mordorintelligence.com/industry-reports/smart-building-market Accessed 18 Feb 2021

[7] http://www.exceedproject.eu/#

OECD (2009), Smart Sensor Networks: Technologies and Applications for Green Growth. Available via http://www.oecd.org/officialdocuments/publicdisplaydocumentpdf/?cote=DSTI/ICCP/IE(2009)4/FINAL&docLanguage=En. Accessed 27 Oct 2020

Oliveira J et al. (2018), Polymer-based smart materials by printing technologies: Improving application and integration, *Additive Manufacturing*, (21), pp. 269–283.

Simonetti R (2017), Why "Smart" Buildings Will Be Standard By 2030. Available via MIPIM World Blog at: https://blog.mipimworld.com/innovation/why-the-smart-building-will-be-standard-by-2030/ Accessed 18 Feb 2021

Talon C, Strother N (2017), White Paper. 10 Trends for Intelligent Buildings in 2017 and Beyond. Digital Transformation and Market Evolution. Navigant Research, 1Q, Boulder (CO), USA

# Chapter 3
# Technology Advancements Changing Architecture

As we saw so far, the building envelope constitutes the element through which buildings establish a "dialogue" between the external environment and the needs of end users; that means it can be considered the "place" in which the various performance issues involved come to a balanced synthesis, thanks to the technological responses of materials and components that made it up.

However, through the incredible modifications that building envelopes suffered over time, due to the changes that occurred in ways of living and human habits as well, we can say that the different technological innovations—that evolved accordingly—have performed a significant role in shaping architecture, leading to the production of buildings with formal and expressive values destined, in turn, to change in a continuously evolving relationship among evolutions and experimentations.

## 3.1 How Technology Is Transforming Architectural Practice

Since the first human life forms have appeared on the Earth, the building envelope constituted a shelter to protect its occupants from external actions of every kind.

A sort of constant among the different forms of dwelling that succeeded over time is the fact that they always tried to exploit positively the environmental context of reference, producing organisms whose formal and technical-constructive features interacted (more or less) coherently with the surroundings.

However, the concept of building envelope, intended as we do right now, takes its origin from Bahnam's (1969) work, in which he introduced the concept of the *well-tempered environment*. Indeed, analyzing the history of architecture, Bahnam recognized three main *environmental control models*, each of them associable with a building envelope type and with specific features.



V. Frighi, *Smart Architecture – A Sustainable Approach for Transparent Building Components Design*, https://doi.org/10.1007/978-3-030-77606-0_3

The first model identified was the *conservative* model, characterized by an environmental control that makes use of massive walls, with lack of openings, to reduce heat losses in a cold climate as well as overheating during hot periods; the second model was the *selective* model suitable for the hot-humid climate because—different from the previous one—it exploits favorable external conditions within confined spaces, such as natural lighting for passive heating. The third and last model instead, was the *regenerative* model, spread where buildings have inadequate mass—being only a barrier among internal and external—thus, the environmental control is entrusted to technological systems.

Such evolution, although with some variations, produced over time several examples of artifacts, able, in a more or less effective way to interact with their application contexts. This has generated a sort of "shared approach" to exploit positive common features, reducing the impact of negative factors to determine efficient building behavior.

The design choices that resulted from this approach, coupled with the evolution and modification of structural systems determined by progress' advancements, shaped in history certain distinctive features of the architecture. So, if early building envelopes were formed on the main features of available materials[1] and technologies, progressively, the production of new tools, the discovering of new materials, and the improvement in construction techniques led them to evolve in a direct relationship with housing conditions, allowing to recognize distinct building elements or techniques as characteristic of a precise point in time, being significant examples of the period that generated it.

In addition to such representative buildings, even traditional houses are in a certain way significant of the characteristics of the place in which they were erected, of the climate, and of the resources available, thanks to the design strategies adopted as well as to the materials employed according to the approach explained before (Fig. 3.1).

However, it is only the advancement in the technological progress that characterized the years between the nineteenth and twentieth centuries when architectural languages underwent a real revolution due to the introduction of new industrial processing techniques and materials (e.g., reinforced concrete, glass, and steel) that allowed the realization of ever more light and transparent casings, in contrast to ancient and strongly massive load-bearing structures. Especially, in the post-war period, progresses in material science promoted limitless possibilities that profoundly altered the nature of architectural design. Following this enthusiastic tendency, architecture became universal, detached from a precise application context, and inspired only by trends and materials as well as by possibilities of technology.

Besides this "flattening" of the architectural features, new requirements have been flanked to ancient primary needs due to new emerging ways of life.

[1] First attempts to operate an environmental control through building structures were tents or cave-shaped shelters which employed timber frame structures with bamboo frames, cladded with leaves or textiles, sometimes equipped with a symbolic meaning and generally with provisional features suitable for nomadic ways of live.

Fig. 3.1 (a) (on the left) Traditional massive buildings. Photo credit: the Author. (b) (on the right) Elements Exhibition Dom-ino Installation—Venice Architecture Biennale 2014. Source https://www.flickr.com/photos/darrel_ronald/14418551557 by Darrel Ronald (own work). CC BY-NC-ND 2.0 (https://creativecommons.org/licenses/by-nc-nd/2.0/)

This has resulted in an acceleration of the aging of a great number of recent buildings, making them inadequate toward specific environmental and performance requests. This is the reason why, to overcome this issue, robust use of technological systems began, making the model so unsustainable in a short time.

Thanks to the awareness that this development would not be possible anymore, coupled with the understanding of the importance that the implementation of sustainable building practices would have, another change in the understanding of the building's envelope occurred. Indeed, it started to be conceived as a set of layers, each of them equipped with a different task, able to perform precise functions and to meet current performance requirements. Building parts previously made of one single material were now engineered and assembled using several different sources and suppliers, privileging materials light in weight and thin in section (Trubiano 2013). However, the global fascination that led to the construction of skins lighter and thinner than ever before has resulted sometimes in negative consequences for the energy performance of buildings: Technologically advanced buildings are often the poorest performing and routinely condemned for not addressing the energy required to maintain their interiors conditions (Trubiano 2013). Also, because of the extreme variation in the external boundary conditions, building envelopes with static features have shortly revealed their limits in intelligently controlling all the parameters responsible for their correspondence to current regulations as well as to the continuously evolving needs of their end users.

Such a situation led professionals and researchers in the field to search for technical solutions able to overcome this shortcoming, favoring a new concept of building envelope to emerge.

Thus, building systems were no more conceived following the conventional relationship between needs and performance—in linear logic of direct proportionality between climatic conditions and requirements of building materials and component—but rather they started to be imagined as objects able to change their performance responses according to the variability in climatic conditions, characterized by the adaptability and the flexibility of their formal and technological structures regarding the external environment.

> "This peculiarity, which is the "adaptive" ability, can be either passive or active; the latter case includes the possibility of changing the conformation of the technical elements to activate, therefore, a proper interaction between outdoor and indoor environment" (Lucarelli et al. 2012).

In general terms, indeed, it can be said that there are two levels of mechanisms that drive the adaptive behavior: the *microlevel*, which implies changes in a smaller scale (general within the material itself), and the *macro-level*, often associated with motions of various kinds. In macrolevel adaptive behavior, the driving principle is generally an external energy input. Despite this subdivision, combinations are also seen.

Such possibilities, previously unthinkable, led to a further change in how designers think and conceive building envelopes, influencing the development of modern architectures whose designs are distinguished by increasingly complex forms. In this innovation process, it is a clear fact that even architectural language suffered significant modifications due to the new opportunities introduced in this field that radically changed how we think and build.

## 3.2 Reconciling the Multiple Performance in a Holistic Way: The CABS

The complexity of the current society coupled with the evolution of building envelope performance requirements—continuously implemented by new needs and efforts aimed at bridging the gap between current building practice and future energy-saving targets—has led the building envelope to evolve into a system different from the previous reference model, based on static designs that do not possess the flexibility to adapt and react to changing conditions.

In the early past, indeed, "conventional" building envelopes typically had static properties (suitable, in their average values, for the near surroundings in which they stand), the fixed ratio among transparent and opaque surfaces, fixed thermal and acoustic insulation values, and—sometimes—a great amount of glazed surface, often not appropriately shaded. In general, their design was scarcely optimized for energy use, transferring the control of the internal comfort condition to massive use of energy for heating and cooling.

However, even if formerly this approach sounds rather functioning, nowadays it no longer seems possible; for these reasons, "modern" building envelopes started to be conceived as *dynamic systems*, designed in a focused and prompt way to offer variable performance according to the variations in environmental conditions, following a nonlinear approach.

In the aim to design architectural built environments that better meet users' needs, adaptive design strategies started to be developed, holding great promises for the next generation of high-performance buildings.

So, to the three typologies of building envelope seen before (see Sect. 3.1), a fourth model could be added, called *smart building envelope*. Frequently defined in the literature as Climate Adaptive Building Shell[2] (CABS) or with other complementary terms,[3] these systems are equipped with unconventional features able to move building performance beyond the level of the best static building envelope design.

The idea to develop adaptive and responsive buildings finds its origin in the early 1980s when Mike Davies (1981) proposed[4] the concept of a polyvalent wall (Fig. 3.2); furthermore, this conception has been alternatively chased over years, finding inspirations even from nature's adaptability as survival strategies.

Such dynamic systems "can be compared to a living organism, in which each part react to external and internal stimuli, adapting to the surrounding context to regulate and optimize the overall energy balance necessary for its functioning" (Romano et al. 2018) (Table 3.1).

Although different types of adaptive facade concepts have been developed over the years, their implementation is still in the beginning stage; indeed, there is no

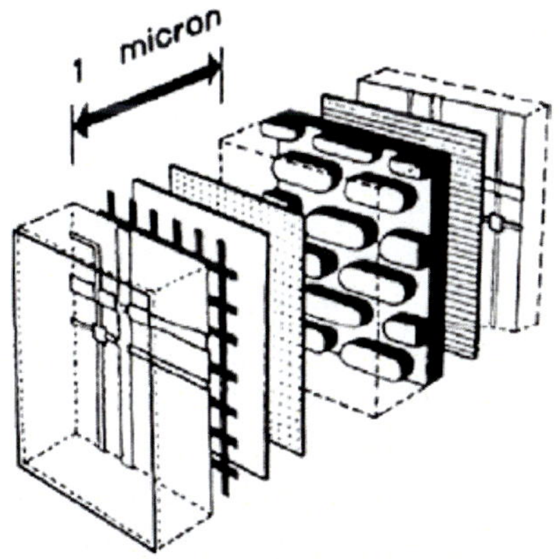


**Fig. 3.2** Schematization of the concept of Davies' polyvalent wall. Source: AM Envelope. The potential of Additive Manufacturing for facade constructions by Holger Strauss (Own work). CC BY 4.0 (https://creativecommons.org/licenses/by/4.0/deed.en) via Research Gate (https://www.researchgate.net/publication/307823959_AM_Envelope_The_potential_of_Additive_Manufacturing_for_facade_constructions). https://doi.org/10.7480/abe.2013.1

[2] Firstly by Loonen (2010), who set, we can say, the first state-of-the-art on the topic.

[3] Such as switchable (SWIFT 2003), adaptive, responsive (Kirkegaard and Foged 2011), dynamic (Lollini et al. 2010), intelligent (Stacey 2011; Wigginton and Harris 2002; Ochoa and Capeluto 2009), reactive, interactive (Fox and Kemp 2009), kinetic (Fox and Yeh 1999), sensitive, and many other synonyms used in an interchangeably manner.

[4] In his work Davies presented the idea of a generic building envelope, not limited to a particular building type, that would perform different complementary and simultaneously actions such as nano-metric absorber, radiator, reflector, filter, etc.

**Table 3.1** Analogy between building needs and biological systems

| Biological system | Adaptive action performed | Adaptability feature | Similar building component | Corresponding Smart material |
|---|---|---|---|---|
| Insects | Optimization of lighting | Size, location, color, and efficacy | Technical systems | LEDs, Electroluminescent, Chemiluminescent paints |
| Spiders and scorpions | Monitoring of structural systems | Stress and deformation, crack and vibration monitoring | Self-healing/ repair devices for monitoring facades | Fiber-optics, Piezo-electrics |
| Lotus leaves | Surface finishes | Self-cleanings, heat and radiation reflection, durability | Coatings | Self-cleaning paints and finishes, conductive and luminesce paints |

consistent definition of facade adaptability even if studies related to characterization issues, design parameters, and classification already exist.

According to literature, there are more than 30 definitions for an intelligent building, intended as a *technical enhanced building*, even if a recurrent statement defines such systems as those able "to change some of their functions, features or behavior over time, in response to changing performance requirements and variable boundary conditions [...] with the final goal of improving overall building performance" (Loonen et al. 2013).

Knaack and Klein (2008), again, supports the use of the term *adaptive* as an alternative of *intelligent* or *smart*, because, as he stated, "even if buildings able to adapt to changing climatic conditions are called intelligent buildings, [...] the term intelligent can be misleading when used in the context of buildings or façades, [so] we will use the term adaptive façade instead".

An intelligent building indeed is an "object" designed to "employ technologies to get the best advantage, in order to make the most of the resources and facilities available" (McClelland 1988), defined by the Intelligent Building Institute of Washington, D.C., as something "which integrates various systems to effectively manage resources [...] to maximize: occupant performance, investment and operating cost savings and flexibility" or, in other words, "that incorporates the best available concepts, materials, systems and technologies [...] to achieve a building which meets or exceeds performance requirements of buildings' stakeholders" (European Intelligent Building Group).

Therefore, the difference between "intelligent" and "smart" is implicitly related to the type of system's control: In the first case, automation is used as controlling elements while in the second case, control is entrusted by properties of the materials.

Regarding other terms previously mentioned, each of them could be used and intended with slightly different meanings: "Interactive," for instance, could imply the use of technology but still in combination with human inputs (Velikov and Thun 2012).

So, despite the multiple available definitions, it is interesting to consider *smart buildings* as an evolution of nearly zero energy buildings, in which the possibility of dynamically controlling the building's environmental parameters, according to the variation in external or internal conditions, is added.

Therefore, it is evident fact that a shared definition, in the literature, is still missing, as each author emphasizes certain issues over others. Addington and Schodek, already in 2005, stated that "the blurring of boundaries between disciplines has given rise to a near crisis in the definition of their respective roles. […] Even within the discipline of architecture, terms such as "smart", "responsive" or "adaptive" have been used loosely and interchangeably, creating confusion as to their specific meaning and their conceptual relationship to building performance and design" (Fig. 3.3; (Table 3.2)).

However, coherently, with the purposes of the present work, a general definition is provided here, trying to recap and synthesize what are the key aspects in facades' intelligence and adaptability.

| Terms | Definition |
|---|---|
| Kinematics & Kinetics | Kinematics is the study of motion through system geometry without considering mass or external forces that may cause movement (RAO et al., 2003; Meriam and Kraige, 2012).<br>Kinetics is the study of movement laws, including forces and masses, and emphasizes on the relationship between motion and its causes (Lienhard, 2014). |
| Dynamics | Defined as the study of movement which occurs due to forces applied to an object (Meriam and Kraige, 2012). |
| Retractable | Term commonly used in architecture for defining textile membrane roofs, either bunched or folded, a term or expression similar to movable (Barozzi et al., 2016). |
| Convertible | Mostly associated with building mobility, similar to building envelopes, that suggest change form in the short-term depending on needs and functions (Otto, 1971). |
| Transformable | The term describes units or systems with the intrinsic property of controlled change, such as the ability to be foldable, retractable or shape-shifting (Hoberman, 2006). |
| Performative | Referred to systems that can mediate surrounding environment for user comfort, such as building skin systems able to control external factors in relation to predefined architectural performances (Turrin et al., 2012). |
| Interact | The term refers to either intuitive or automatic responses to meet user requirements. Interactive building architectures use sensors to operate kinetic systems and enable intelligent materials in order to initiate change in appearance and the environment (Sung, 2012). |
| Smart | Process-related concept that conveys automatic control or operation of building envelopes, including heat loss or gain, daylighting, air ventilation and other systems. Smart systems use information technology to connect several subsystems that typically operate independently (Brooks, 2011). |
| Responsive | Referred to a system moving and responding from the outside based on specific factors, thus allowing interaction with a passive environment (Hasselaar, 2006). |
| Adaptive | The ability of a system to adjust by itself in relation to a changing environment. An adaptive system, as in the case of building skins, has the ability to adapt the features, behavior or configuration of the external environment (Dewidar et al., 2013). |

**Fig. 3.3** Summary of common terms employed in contemporary architecture to define its "adaptability" feature. For further insights, refer to Al-Obaidi et al. (2017)

**Table 3.2** Synoptic comparison of terms recently employed to alternatively defined the CABS and related references

| Terms | Reference |
|---|---|
| Smart | Fox and Yeh (1999)<br>Velikov and Thun (2012)<br>Brugnaro et al. (2014)<br>Casini (2016) |
| Switchable | Beevor (2010) |
| Adaptive | De Marco Werner (2013)<br>Möller and Nungesser (2015) |
| Responsive | Negroponte (1976)<br>Ferguson et al. (2007)<br>Kirkegaard and Foged (2011)<br>Heiselberg et al. (2012)<br>Kolodziej and Rak (2013)<br>Velikov and Thun (2012)<br>Meagher (2015) |
| Dynamic | Lollini et al. (2010) |
| Intelligent | Hayes-Roth (1995)<br>Kroner (1997)<br>Compagno (2002)<br>Wigginton and Harris (2002)<br>Clements-Croome (2004)<br>Knaack and Klein (2008)<br>Ochoa and Capeluto (2009)<br>Stacey (2011)<br>Velikov and Thun (2012)<br>Masri (2015) |
| Interactive | Fox and Kemp (2009)<br>Velikov and Thun (2012) |
| Kinetic | Fox and Yeh (1999)<br>Loonen (2010)<br>Ramzy and Fayed (2011)<br>Wang et al. (2012)<br>De Marco Werner (2013)<br>Fortmeyer and Linn (2014) |
| Active | Ochoa and Capeluto (2008) |
| Advanced | Ad et al. (2011) |
| Biomimetic | Vermillion (2002) |
| Bio-inspired | Loonen et al. (2015) |
| Controllable | Konstantoglou et al. (2013) |
| Movable | Schumacher et al. (2010) |

[a smart/advanced/adaptive building is] a construction in which different technologies and devices (sensors, readers, computers, etc.) are allowed to interact within the building envelope—among them and with the surrounding—to provide unconventional behaviors, or to collect data or information useful for further implementations.

Such definition takes its origin from the performance that architecture must ensure to appropriately respond to actual standards through its envelope, thus, to tailor its behavior to better meet all the requirements, which basically can be resumed as follows:

- heat insulation control;
- acoustic control;
- sun protection, during hot seasons;
- daylight control and active shading, when needed;
- maximization of SHG, during cold seasons;
- active air and ventilation control;
- energy production, through renewable sources;
- indoor air quality control;
- passive design strategies and exploitation of the site's resources;
- reports and data on real performance, during operation time; and
- self-adjustment capabilities based on users' feedbacks or predetermined input sceneries (Fig. 3.4).

Therefore, a smart adaptive building envelope must:

- *adjust* its performance—according to hourly, diurnal, or seasonal variations—to reduce the impact of external changes in the environment that can be of different orders of time. The building envelope is first influenced by short-term fluctuations that change in the order of seconds, such as wind speed or wind directions; second, by changing conditions in the order of minutes, such as cloud variations or daylight availability. Additionally, changes in the order of magnitude of hours and/or diurnal changes are present as a consequence of the occupants' behavior and/or meteorological boundary conditions. Finally, the last group of variations is seasonal changes during winter, summer, spring, and autumn due to different boundary conditions, such as sun altitude.
- *adapt* itself to the abovementioned variations at the macro-scale (that means through the envelope in its whole) or at the micro-scale (that means through a

**Fig. 3.4** Cannon Design (1981), The Occidental Chemical Building, already known as the Hooker Building at Niagara Falls, New York. It is one of the first buildings that incorporate intelligent responses in its skin; the protecting louvers indeed tilt automatically to keep the sun's direct beams from striking the internal glass wall, thanks to the action of a solar cell on each bank's louver. (Courtesy of Terri Meyer Boake)

single material). According to Loonen et al. (2013), macro-level adaptability is related to the mechanical movement of an envelope's parts, resulting from sliding, expanding, creasing, rolling, inflating, and so on. This can be obtained with external, supplemental components or building envelope's components, or, again, by the movement of the entire facade or the building as a whole, while micro-scale adaptability mainly attains to materials' properties changes (thermos-physical, optical, energetic, etc.) in response to different variations.

- *control* its modes of operation in an extrinsic (that means they make use of feedback to implement their operation using sensors, processors, or actuators with the local processor or through a control unit) or intrinsic (that means they self-adjust their behavior according to external stimuli) way.

Hence, this allows to draw some significant agreements about adaptive facades in general, conceiving them as systems able to provide an adequate response to changes in the internal and external environment, ensuring or improving functional requirements of building envelope in terms of controllable insulation and thermal mass, radiant heat exchange, ventilation, energy harvesting, daylighting, solar shading, or humidity control, in few words able to deliver intended functionality (Figs. 3.5 and 3.6).

**Fig. 3.5** Widely heralded as "the world's first intelligent building," the edifice designed by Skidmore Owings and Merril in Hartford, Connecticut, employs an integrated services system to manage and link together the previously separated functions of air-conditioning, mechanical controls of vertical connections, and safety systems. Source: (**a**) https://commons.wikimedia.org/wiki/File:City_Place_I,_Hartford,_CT.jpg by Xnatedawgx (own work). CC-BY-2.0 (https://creativecommons.org/licenses/by/2.0/deed.en) and (**b**) https://commons.wikimedia.org/wiki/File:Hartford,_Connecticut_-_City_Place_and_Goodwin_Square_(0669).jpg by GeorgHH (own work) GFDL (http://www.gnu.org/copyleft/fdl.html), via Wikimedia Commons

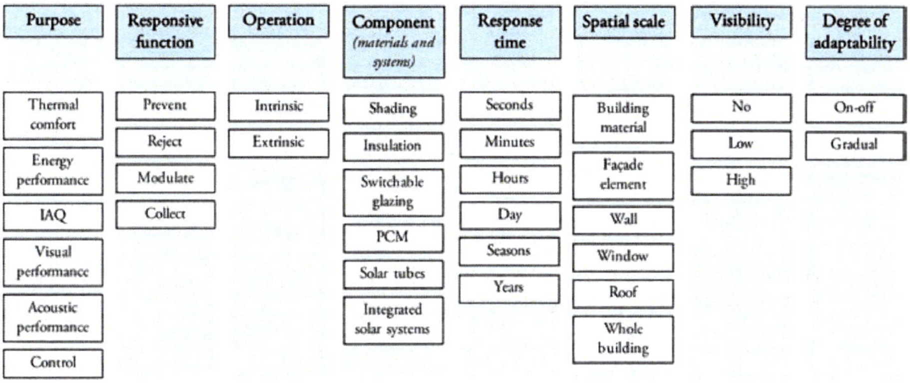


**Fig. 3.6** Purposes for adopting adaptive functions in the building envelope. For further insights, refer to Aelenei et al. (2016)

## 3.3 Smart Advanced Technologies: Innovation and Experimentation

Coherent with what was stated so far, and despite the growing interest toward smart building envelopes, it is quite evident that a common ground in which innovative materials and technologies currently available stand is still not existing, even if different efforts have been made in that direction.

The current situation is made of dispersed and individual research activities, too often lacking in cooperation with productive and industry field, which mainly deals with specific performance evaluation of dynamic insulation or smart windows, just to mention a few.

Despite these attempts, it remains unclear what type of desired behavior results from their applications and what are the best technical solutions as well as how the available technologies can be profitably integrated within building systems.

As an attempt to give an overview of smart advanced technologies suitable for the present work, a classification that descends from families of building envelopes' requirements previously identified (see Chap. 1) has been done.

Advanced technological solutions identified, among those that show smart characteristics, have been classified according to the following categories (Fig. 3.7):

- *advanced, integrated facades*, able to provide high performance as a whole, both actively and passively activated;
- *adaptive glazing* or *daylighting systems* in general, able to regulate the amount of incident solar radiation over time to avoid overheating during summer while maximizing gains in winter;
- *thermal massive* components, employed with storage purposes, and
- *dynamic insulation systems*, able to face seasonal variations. Summer days, for instance, need low insulation levels while during winter, thermal insulation is important to keep heat inside the environment.

**Fig. 3.7** Smart buildings' advanced technologies and their relation to climatic conditions. For further insights, refer to Casini (2016)

| Climate | Smart Envelope materials and technologies |
|---|---|
| HOT | • Low SHGC glazing or dynamic glazing<br>• Dynamic shading systems<br>• BIPV and PV glazing<br>• Cool roofs and reflective technologies<br>• Phase Change Materials (thermal inertia)<br>• Green roofs and walls<br>• Self-cleaning glazing and surfaces<br>• Ventilated roof and walls<br>• Adaptive skin facades |
| COLD | • Advanced insulation glazing<br>• Glazed double-skin facades<br>• Dynamic shading systems<br>• Antireflective glazing<br>• Light redirection and optical systems<br>• Heating glazing<br>• Bio-adaptive glazing<br>• ETFE envelope<br>• BIPV technologies<br>• Self-cleaning glazing and surfaces<br>• TIM<br>• Phase Change Materials (passive solar gains)<br>• Advanced insulating materials<br>• Adaptive skin facades |

The first category comprises all the systems defined as an *advanced facade*, which means double skin surfaces as well as the composition of materials and components able to respond as a whole to particular needs, both statically and dynamically. Within this category, further subdivision can be done according to facades modes of operation whether they are initiated by external stimuli (such as temperature variations), through automated control, or by mechanical (kinetic abilities) of certain components.

*Adaptive glazing* (or daylight systems in general) instead attains all the technologies related to selective control of the incoming incident solar radiation, aimed at reducing, controlling, and filtering its entry within rooms. They can be systems with excellent static performance—that do not need activation—selective (active) components, or chromogenic devices, which are systems able to vary their visible appearance, thanks to different external inputs. In this category fit even solar control louvers or shading system in general, especially in their advanced or self-regulating form.

*Thermal massive components* are all the devices that exploit thermal inertia (and eventually variable thermal mass) "to attenuate peak loads and keep temperature fluctuations within the comfortable range" (Hoes et al. 2011).

In the end, we found *dynamic insulation systems* sometimes introduced in windows systems (by putting insulation materials within shutters) or in the light opaque wall (between the different layers) to prevent heat losses in winter. Recently, different control strategies have been adopted to operate an adaptive insulation level, generally based on the control among heating and cooling loads between outside and inside. Within this category, one of the most promising existing technologies is PCMs, used to store thermal energy while exploiting phase change variations. Indeed, thanks to their ability to store and further release heat, they can reduce peak loads in buildings, increasing the thermal inertia of the component in which they are embedded.

Since the energy storage capacity of PCMs is greater than in common building materials, they are often used in lightweight construction, whose adoption allows for better control of thermal flows.

However, it is generally said that smart features in buildings are often resulting from physical interactions among different components and their related domain (Fig. 3.8).[5]

While most of the abovementioned solutions generally contribute to energy and comfort optimization, examples for adaptation with completely different intentions

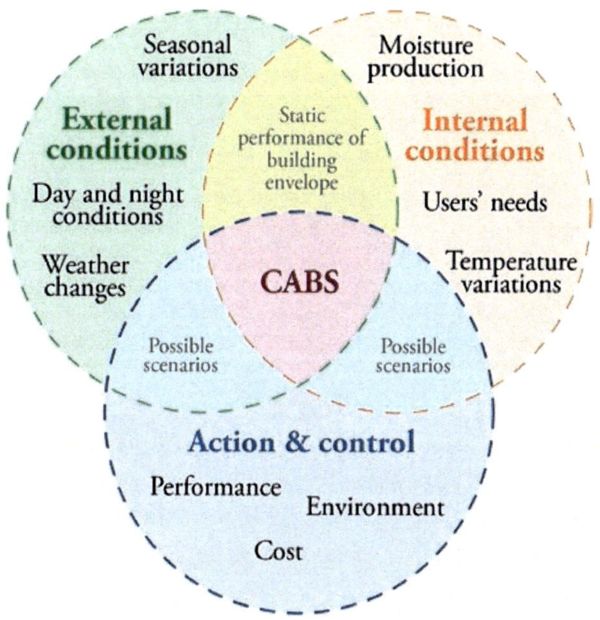


**Fig. 3.8** Domains of CABS: Their interaction can origin 15 different possible combinations. For further insights, refer to Casini (2016)

[5] Loonen (2010) recognized four domains for classify building systems, according to the physical interactions they affect, that are: (1) thermal, (2) optical, (3) air flow, and (4) electrical. CABS that affects the first domain generally cause changes in the energy balance of the building though processes of heat exchange per conduction, convection or radiation or by heat-storage processes; second domain systems are related to visual perception within confined spaces, that changes due to surface modifications in terms of transparency; third domain systems are related to ventilation effect near the building systems (generated by the façade or by effects of wind direction or speed) while the last domain refers mostly to the conversion of energy.

are also found. Well-known cases are the *media facades* that target aesthetic expressions, art, or commercial media performance, most often not incorporating energy use aspects.

Thus, the adaptation of facades seen from the perspective of architecture is not necessarily linked with performance optimization; even if newer developments of media facades show, even more often, the employment of renewable energy for its operation or consider their elements as multifunctional components that contribute to the energy concept (Fig. 3.9).

Within the field of smart and advanced technologies, a special mention is deserved by *sensors* and *actuators*, fundamental components of smart systems. Indeed, too often, we tend to merge the technologies employed with their operational controls even if they are quite commonly separated.

Control is part of a system that includes a sensor, an active device, an interface for inputting the desired response, and an algorithm for controlling the response of the device. So when a system is activated, thus fulfilling the expected goal, an input able to generate its reaction is needed. It can be external or internal, defining, respectively, *extrinsic* or *intrinsic controls*.

*Extrinsic controls* are those based on sensors, actuators, and processors: Sensors "connect" the adaptive system with the external environment, transmitting the

**Fig. 3.9** Chemnitz's (DE) main station interactive LED facade, which incorporates the movements of the surrounding and projects them as light movements on the building envelope. Architect: Grüntuch Ernst Architekten. Light Art: Random International, London. Source: https://commons.wikimedia.org/wiki/File:Chemnitz_main_station_with_installation_at_media_facade_by_Simon_Hillme.jpg by Alice d25 (own work). CC BY-SA 4.0 (https://creativecommons.org/licenses/by-sa/4.0/deed.en)

variations in reference parameters' changes and, after confronting them with the set-points, transferring such information to the processor; actuators are the elements that produce a reaction in the system following a signal received from sensors and then elaborated by a processor; and processors are the control units in which collected data are elaborated to provide an adequate response, according to design provisions.

On the contrary, *intrinsic controls* (even defined as passive systems) are those that can count on the ability of the systems to self-adjusting in response to external stimuli, generally of environmental nature. The most relevant expressions of such definition are the so-called Smart Materials, which provide a significant evolution in the domain of the building envelope.

As researchers in the field of architecture, by better understanding how smart technologies and material systems work, architecture can be pushed forward to truly reach a new era of innovation, creating better and more adaptive built environments with optimal consequences for their users.

## References

Ad A, Heiselberg P, Perino M, (2011), Designing with responsive buildings elements. IEA - ECBCS Annex 44, Aalborg University

Addington M, Schodek D (2005), Smart Materials and New Technologies. For architecture and design professions. Architectural Press, Elsevier, Oxford (UK)

Aelenei D, Aelenei L, Pacheco Vieira C (2016), Adaptive Façade: Concept, Applications, Research Questions. Energy Procedia, 91, 269-275

Al-Obaidi KM, Ismail MA, Hussein H, Rahman AMA (2017), Biomimetic building skins: an adaptive approach. Renewable and Sustainable Energy Reviews, 79, 1472-1491

Bahnam R (1969), The Architecture of the Well-Tempered Environment. Architectural Press, Elsevier, London

Beevor M (2010), Smart Building Envelopes, 4th Year Project Report. University of Cambridge, Department of Engineering.

Brugnaro G, Caini M, Paparella R (2014), Energy saving through Building Envelope Innovation: Smart Skin Design. Recent Advances in Urban Planning. Sustainable Development and Green Energy, pp 35-44

Casini M (2016), Smart Buildings. Advanced Materials and Nanotechnology to Improve Energy-Efficiency and Environmental Performance. Woodhead Publishing Series in Civil and Structural Engineering: Number 69, Elsevier, Oxford (UK)

Clements-Croome DJ (2004), Building environment, architecture and people. In: Clements-Croome DJ (ed.), Intelligent buildings: Design, Management and Operation. Thomas Telford, London, pp 53-100

Compagno A (2002), Intelligente Glasfassaden: Material, Anwendung, Gestaltung. Birkhauser, Basel

Davies M (1981), A wall for all seasons, RIBA Journal, 88(2):55-57

De Marco Werner C (2013), Transformable and transportable architecture: analysis of buildings components and strategies for project design. Dissertation, Escuela Técnica Superior de Arquitectura de Barcelona, Barcelona

Ferguson S, Siddiqi A, Lewis K, De Weck O (2007), Flexible and reconfigurable systems: Nomenclature and review. In: ASME 2007 International Design Engineering Technical

Conferences and Computers and Information in Engineering Conference, Las Vegas, Nevada, USA, pp 249-263

Fortmeyer R, Linn CD (2014), Kinetic Architecture: design for Active Envelope. The Images Publishing Group Pty Ltd., Australia

Fox M, Kemp M (2009), Interactive architecture. Princeton Architectural Press, New York

Fox MA, Yeh BP (1999), Intelligent kinetic systems in architecture. In: Proceedings of managing interactions in smart environments, MANSE'99, Dublin, 1999, pp 91-103

Hayes-Roth B (1995), An architecture for adaptive intelligent systems. Artificial Intelligence, 72 (1–2):329-365

Heiselberg P, Inger A, Perino M (2012), Integrating Environmentally Responsive Elements in Buildings, In: Proceedings of the 27th AIVC Conference, Lyon, 2012

Hoes P, Trcka M, Hensen J (2011), Investigating the potential of a novel low-energy house concept with hybrid adaptable thermal storage. Energy Conversion and Management, 52(6):2442-2447

Kirkegaard PH, Foged IW (2011), Development and evaluation of a responsive building envelope. In: Proceedings of adaptive architecture, London, 2011

Knaack U, Klein T (2008), The Future Envelope 1: A Multidisciplinary Approach. IOS Press, Amsterdam

Kolodziej P, Rak J (2013), Responsive building envelope as a material system of autonomous agent. In: Open Systems: Proceedings of the 18th International Conference on Computer-Aided Architectural Design Research in Asia, pp 945-954

Konstantoglou M, Kontadakis A, Tsangrassoulis A (2013), Dynamic Building Skins: Performance Criteria Integration. Paper presented at the PLEA2013 - 29th Conference, Sustainable Architecture for a Renewable Future, Munich, 2013

Kroner WM (1997), An intelligent and responsive architecture. Automation in Construction, 6, pp 381-393

Lollini R, Danza L, Meroni I (2010), Energy efficiency of a dynamic glazing system, Solar Energy, 84(4):526–37

Loonen RCGM (2010), CABS – What can we simulate? Dissertation, Eindhoven University of Technology

Loonen RCGM, Rico-Martinez JM, Favoino F et al (2015), Design for façade adaptability – Towards a unified and systematic characterization. In: Proceedings of the 10th Conference on Advanced Building Skins, Bern, Switzerland, pp 1284-1294

Loonen RCGM, Trčka M, Cóstola D, Hensen JLM (2013), Climate adaptive building shells: State of the art and future challenges, Renewable and Sustainable Energy Reviews, 25, pp 483–493.

Lucarelli MT, Mandaglio M, Pennestrì D (2012), The building envelope between process and product innovation. Paper presented at the International Conference: Sustainable Environment in the Mediterranean Region: from Housing to Urban and Land Scale Construction, Naples, Italy, 12-14 February 2012

Masri Y (2015). Intelligent Building Envelopes: Design and Applications. In: Proceedings of the International Conference on Building Envelope Design and Technology, Graz Advanced Building Skins 2015, pp 37-46

McClelland S (1988) (ed), Intelligent Buildings; An IFS Executive Briefing. Bedford, England, IFS Publication/Springer Verlag, Blenheim Online.

Meagher M (2015), Designing for change: The poetic potential of responsive architecture. Frontiers of Architectural Research, 4, pp 159-165

Möller E, Nungesser H (2015), Adaptable Architecture by Frei Otto: a case study on the future viability of his visions and some forward ideas. In: Proceedings of the International Association for Shell and Spatial Structures, Amsterdam, 2015

Negroponte N (1976), Soft Architecture Machines. MIT Press, Cambridge

Ochoa CE, Capeluto IG (2008), Strategic decision-making for intelligent buildings: Comparative impact of passive design strategies and active features in a hot climate. Building and Environment, 43 (11):1829-1839

Ochoa CE, Capeluto IG (2009), Advice tool for early design stages of intelligent facades based on energy and visual comfort approach, Energy and Buildings 41(5):480–8
Ramzy N, Fayed H (2011), Kinetic systems in architecture: New approach for environmental control systems and context-sensitive buildings. Sustainable Cities and Society, 1(3):170-177
Romano R, Aelenei L, Aelenei D, Mazzucchelli ES (2018), What is an Adaptive Façade? – Analysis of Recent Terms and Definitions from an International Perspective. In: Facade 2018 - Adaptive!, Lucerne, 2018
Schumacher M, Schaeffer O, Voght MM (2010), Move. Architecture in Motion – Dynamic Components and Elements. Birkhauser, Basel
Stacey M (2011), Adaptive architecture introduction. In: Proceedings of adaptive architecture, London, 2011
EU-SWIFT (2003), Architectural and Technical Guidelines - Handbook for the Use of Switchable Facades Technology
Trubiano F (2013) *Design and Construction of High Performance Homes*, Routledge Press: London
Velikov K, Thun G (2012), Responsive Building Envelopes: Characteristics and Evolving Paradigms. In: Trubiano F (ed) Design and Construction of High Performance Homes. Routledge Press, London, pp 75–92
Vermillion J (2002), Phototropic architecture: intelligent responses to sunlight stimuli. A new school of architecture for Ogerlthorpe University, Atlanta. Dissertation, Ball State University
Wang J, Beltrán L, Kim J (2012), From Static to Kinetic: A Review of Acclimated Kinetic Building Envelopes. In: Proceedings of The Solar Conference, 5, pp 4022–2029.
Wigginton M, Harris J (2002), Intelligent skins, Butterworth-Heinemann, Oxford

# Chapter 4
# What Is a Smart Architecture?

## 4.1 Smart Architecture: Recent Developments and Examples

According to what has been stated in previous chapters, innovative smart advanced technologies (referred to the specific domain of building envelope), under their abilities in implementing buildings' features, seem equipped with great potential, not only from an aesthetic point of view—due to the almost unlimited formal possibilities that they allow—but even from a performance-oriented perspective, thanks to the opportunities they offer in controlling building energy consumption, making them suitable for different purposes and needs.

To complete the systematic characterization of the technologies able to provide building envelope with smart features, as well as to understand how their adoption has innovated building practices, by modifying building architectural language, a sample of contemporary architecture selected as case studies has been analyzed.

This review has been useful to better understand the current state of such innovative applications to:

- provide a holistic qualitative evaluation framework that helps to gain insight into the degree of fulfillment of every building envelope requirement
- develop a map of adopted strategies in contemporary smart buildings
- recognize common patterns and, at the same time, identify unexplored concept to help further the development of high-potential innovative smart building technologies
- investigate the specific role that transparent building components play within smart buildings, understanding their application, technical, and performance potential, thus promoting their full comprehension.



V. Frighi, *Smart Architecture – A Sustainable Approach for Transparent Building Components Design*, https://doi.org/10.1007/978-3-030-77606-0_4

### *4.1.1 Selection*

The sample of smart architecture presented includes approximately 50 buildings (and/or prototypes ascribable to the category of smart buildings as intended in the wide meaning given to this term in the present work). The case studies have been selected among recent architectural expressions that show particular forms of innovations, according to criteria that include environmental sustainability and energy efficiency.

In particular, a huge sample of buildings has been investigated, analyzing their technical characteristics and their conditions of use thus seeking after their "smart feature."

To reach for it, the following have been analyzed: (1) the opening systems and the potential adoption of particular techniques or technologies designed to resolve critical issues related to thermal or visual comfort; (2) the presence of shading devices equipped with particular advanced, innovative, or anyway, uncommon features, thus, for this reason, considered as *smart*; (3) the presence of innovative forms of thermal insulation or advanced insulating materials; (4) the existence of energy harvesting or generation systems; (5) the typology of building envelope and the possible adoption, within it, of innovative technical solutions or, again, of solution that innovatively reinterpret traditional methods and materials, to allow new functioning models and new performance; and finally, more in general, (6) the presence of advanced technologies as described within the previous chapters.

In this selection, even the geographic position has been taken into consideration to gather data coming from different application contexts.

Clearly, the collection does not intend to be comprehensive of the multitude of realizations ascribable under the abovementioned definition; instead, it would provide a general overview within the current state of the art about smart building technologies, discovering what kind of solutions have been seen through current building practice, understating their strength and weaknesses and trying to comprehend which of them can become new architectural languages by their abilities to solve present and future critical issues.

### *4.1.2 Classification: Assessment Criteria and Definition of Parameters*

The sample outlined has been examined according to the following criteria.

First, a general analysis has been done, reporting for each case study's general information, including its *intended use* (residential/service-industry/specialized), its *state of development* (concept or prototype/real-scale edifice/under construction), and the climate condition of its *geographic position*, defined, in a synthetic way, according to the Koppen Climate Classification.

Definitions of each category are given at the end of the paragraph.

Second, the classification follows two different levels of analysis; the first level aims at identifying the *type of component/system* equipped with smart function(s) (insulation/glazing/shading/opening systems/energy harvesting device/energy generation devices/other) and its *application scale* (element/material/component/façade system)[1] while the second level regards case study behavior and it aims to synthetically describe system's operation mode, providing a qualitative analysis of it, by reporting:

- the (main) *purposes for which the technologies have been adopted* (thermal comfort/visual performance/acoustic comfort/energy management/generation/IAQ/interaction and users' control/other).
- the (main) *smart function that such technologies perform* (collect and storage/control/prevent/reject/allow interaction).
- a categorization of the case study according to the following categories:
  - *High performance.* Defined for all the case studies that show high performance or particular features that do not fit with other categories (e.g., advance insulation).
  - *Selective.* Defined for all the case studies that present particular selective filter abilities toward external factors, such as, for instance, blinds or shading systems. This category mainly pertains to materials and components with selective properties toward light and solar control.
  - *Dynamic façade.* Defined for all the case studies with one or more façades able to vary their performance features over time. This has been considered as a separate category because previous ones can be prerogative of a single material or component while this section is intended only for façades as a whole.
  - *Adaptive glazing.* This category can be considered a specialization of the previous one, although, for the specific role that transparent building components play in the present work, it has been separated; within it fit all chromogenic technologies applied to glazed components.
  - *Energy generating.* Defined for all the case studies with particular smart features that allow to integrate within their envelope energy-generation abilities.
  - *Control system.* A category that includes, in a wide sense, all the case studies that allow interaction and users' control.
- the *operation mode* of the smart element (passive, if activated by means of external stimuli of various nature, or active, if activated through electrical control)
- the *type of trigger* of the smart function (thermal/electromagnetic/optical/air quality/occupants/other)
- the *type of response input*, which describes how smart components are controlled (extrinsic/intrinsic/self-adjusting/users' control)
- the *degree of adaptability* that expresses how the component/system with smart functions responds to changing boundary conditions (on-off/gradual).

---

[1] A component is defined as an assembly of different set of elements; it forms a complete constructional or functional unit as part of a facade, while a facade system is intended as a set of different transparent of opaque components; it fulfils all basic technical facade functions such as insulation, rain, and wind tightness.

The last information provided regarding further evaluations aimed at giving a definition of the *technological readiness level*[2] of the technologies identified, intending, in this work, their market spreading application.

Furthermore, a brief description of each case study has been provided, equipped with at least a picture of it. Specific highlights, if available, have been provided.

Main references to the information collected and employed to describe the case studies are reported at the end of each form (Table 4.1).

### 4.1.3 Smart Architecture Examples

Here are listed, in chronological order, the Smart Architectures indexed in the sample investigated:

- Gemini House (1991), Roland Mösl, Weiz (A)
- Heliotrop (1994), Rolf Disch, Freiburg (DE)
- Museum of paper art (2002), Shigeru Ban Engineers, Shizuoaka (JP)
- Lumenhaus (2002), Virginia Tech team, Solar Decathlon competition of Madrid (SP)
- Esplanade Complex (2002), DP Architects, James Stirling and Michael Wilford, Singapore
- Jubilee Church, la Chiesa del Dio Padre Misericordioso (2003), Richard Meier, Rome (IT)
- EWE Arena (2003), wh-p Ingenieure, Oldenburg (DE)
- La Defense Almere (2004), UNStudio, Almere (NL)
- Chanel Ginza (2004), Peter Marino Architect, Tokyo (JP)
- Selimex Offices (2005), Werner Tscholl, Laces, Bozen (IT)
- Kiefer Technic Showroom (2007), Ernst Giselbrecht + Partner, Bad Gleichenberg (A)
- Pola Ginza Building Façade (2009), Nikken Sekkei + Yasuda Atelier, Ginza district, Tokyo (JP)
- North House (2009), Team North, University of Waterloo, Ryerson University and Simon Fraser University, US Department of Energy Solar Decathlon, Washington DC (USA)
- Surry Hills Library and Community Centre (2009), FJMT, Sydney (AUS)
- 100 11th Avenue—Sunset Façade (2010), Jean Nouvel, New York City (USA)
- The breath building, GEOX Megastore (2010), Dante Benini & Partners Architects, Milan (IT)
- ThyssenKrupp Headquarters (2010), JSWD Architekten, Chaix & Morel et Associès, Essen (DE)

[2]The Technology Readiness Level (TRL) expresses the level of technological maturity of a technology. It was initially developed by the NASA in 1974 and further employed by other authorities such as the European Commission (Technology readiness levels (TRL), HORIZON 2020—WORK PROGRAMME 2018–2020 General Annexes, Extract from Part 19—Commission Decision C(2017)7124[3]). It is based on a range from 1 to 9, where 1 is the lower value that express an early technology while 9 is the highest that express a fully developed system.

- Adaptive Solar Façade (2011), Team leaded by Prof. Arno Schlueter, ETH Zurich (CH)
- MEDIA-ICT Building (2011), CLOUD 9—Enric Ruiz-Geli, Barcelona (SP)
- F40 Commercial building (2011), PETERSENARCHITEKTEN, Berlin (DE)
- Quadracci Pavilion, Milwaukee Art Museum (2011), Santiago Calatrava, Milwaukee (USA)
- Kuggen (2011), Wingårdh Arkitektkontor, Gothenburg (S)
- RMIT's Design School (2012), Sean Godsell, Melbourne (AUS)
- Smart Material Houses Project: Smart is green (2012), zillerplus Architekten und Stadtplaner, Hamburg (DE)
- Smart Material Houses Project: Soft House (2012), Kennedy & Violich Architecture, Hamburg (DE)
- South Australian Health and Medical Research Institute—SAHMRI (2012), Woods Bagot, Adelaide (AUS)
- Windswept at Randall Museum (2012), Charles Sowers, San Francisco, CA, (USA)
- One Ocean Thematic Pavilion EXPO (2012), SOMA Lima, Yeosu-Si (ROK)
- Al Bahar Towers (2013), AEDAS UK, Abu Dhabi (UAE)
- Hanwha Headquarters Remodelling (2013), UNStudio, Seoul (ROK)
- Smart Material House BIQ (2013), Arup and Splitterwerk Architects, Hamburg (DE)
- Torre de Especialidades, Hospital Manuel Gea Gonzales (2013), elegant embellishments, Mexico City (MEX)
- Festo's AutomationCenter (2014), Ulrich & Jens Jaschek Architects, Esslingen (DE)
- May-September, Eskenazi Hospital façade (2014), Rob-Ley, Urbana Studio, Indianapolis (USA)
- SDU Campus (2014), Henning Larsen Architects, Kolding (DK)
- The W.I.N.D. House (2014), UNStudio, North-Holland (NL)
- Umbrella façade (2014), Madrid Pavillon at Expo Shangai, 3Gatti Architecture Studio, Shangai (RPC)
- The Breathing Skins (2015), Ing. Tobias Becker, Mandelbachtal (DE)
- Shanghai Natural History Museum (2015), Perkins and Will, Shangai (RPC)
- Forti Holding S.P.A. Headquarters and Office Building (2016), ATIPROJECT, Pisa (IT)
- Siemens Headquarters (2016), Henning Larsen Architects, Munich (DE)
- King Abdullah Petroleum Studies and Research Centre (2017), Zaha Hadid Architects, Riad (ARS)
- Métropole Rouen Normandy (2017), Jacques Ferrier Architecture, Rouen (FR)
- US Embassy (2017), Kieran Timberlake, London (UK)
- The Bund Finance Centre (2017), Heatherwick and Foster + Partners, Shangai (RPC)
- Futurium Berlin (2017), Richter Musikowski, Berlin (DE)

**Table 4.1** Category definition employed to perform the case studies analysis. The table resumes and clarifies the criteria employed to conduct the *general examination* (concerning case study's general information, intended use, state of development, and climatic condition of its application context) and the *two-level analysis* (the first level aims at identifying the type of component/system equipped with smart function(s) and its application scale, while the second level regards case study behavior and the description of its operation mode). Finally, information regarding the technological readiness level of each case study has been provided.

| Category definition | | | |
|---|---|---|---|
| General information | Intended use | Residential | Basic housing type suitable for accommodating one or more family (Caniggia Maffei 1979, p.106) |
| | | Service/industry | Building of the tertiary (service-industry) sector, primarily aimed at host offices |
| | | Specialized | Building commonly suitable for nonresidential use, that means with a high level of specialization (Caniggia Maffei 1979, p.106) |
| First level analysis | Building state of development | Concept/prototype | Pilot building at an early stage of development, intended to be used as a model for further reproduction |
| | | Real scale edifice | Real building |
| | | Under construction | Real building under the construction phase |
| | Climate condition | Tropical | Tropical region with a significant level of precipitations and in which the average temperature of the coldest month is higher than 18 °C; generally, lack of cold season |
| | | Dry/desertic | Region in which the average rainfall level is lower than the dry limit |
| | | Temperate | Warm region in which the average temperature of the coldest month is between 18 °C and −3 °C; lack of a regular snowy coverage |
| | | Continental | Climate with warm summer and cold winter and rich rainfalls over the year |
| | | Polar | Climate with the average temperature of the hottest month lower than 10 °C |
| | Application scale | Element/material | Building material or element manufactured with precise dimensional, geometric, and performance features |
| | | Component | Technical element given by the assembly of different materials or elements with a specific purpose; the term 'component' describes a functional subsystem which contributes to the definition of the façade system |
| | | Façade system | Combination of heterogeneous components, elements and/or materials originating a building envelope system or, at least, a part of it<br>It indicates the entire set of components of a façade, including the control systems that might be used to control adaptive features of it |
| | Type | Insulation | Element mainly aimed at reducing heat flux through the envelope between internal and external environment |
| | | Glazing | Element mainly glazed, thus aimed at ensuring an adequate level of natural lighting and ventilation within confined spaces |
| | | Shading | Element mainly aimed at controlling incident solar radiation |
| | | Opening systems | Element relevant to opening/handling devices, related to both opaque and transparent components |
| | | Energy-harvesting devices | Element aimed at harvest and store energy; even elements/components/systems aimed at increase thermal inertia and frontal mass of building envelope can be inserted within this category |

| | | | |
|---|---|---|---|
| | | Energy-generation devices | Element aimed at producing energy (generally by converting it from one form to another) |
| | | Other | Heterogeneous technologies that do not fit in the abovementioned categories |
| Second level analysis | Purposes | Thermal comfort | Technology/ies adopted to primarily ensure the thermal comfort of the indoor environment |
| | | Visual performance | Technology/ies adopted to primarily ensure an adequate visual performance, which means correctly performing a light/solar control |
| | | Acoustic comfort | Technology/ies adopted to primarily ensure the acoustic comfort of the indoor environment |
| | | Energy-management/ generation | Technology/ies adopted to primarily manage/collect/convert energy |
| | | IAQ | Technology/ies adopted to primarily ensure an adequate air quality level within confined spaces |
| | | Interaction and users' control | Technology/ies adopted to primarily allow the interaction of end users with the building systems |
| | | Other | Technology/ies that does/do not fit in the abovementioned categories |
| | Adaptive function | Collect and storage | Technology/ies that perform a "collect and storage function" as, for instance, regarding heat or energy (e.g., PCM) |
| | | Control | Technology/ies that mainly perform a controlling function regarding different kind of factors, such as, for instance, incident solar radiation |
| | | Prevent/Reject | Technology/ies that mainly perform a preventing/rejecting function towards different kind of factors, such as, for instance toward sound |
| | | Allow interaction | Technology/ies that mainly allow interactions among the building system and its end users |
| | Classification | High performance | Case studies that show high performance or particular features that do not fit with other categories, such as, for instance, advanced insulation |
| | | Selective | Case studies that present particular selective filter abilities toward external factors, such as, for instance, blinds or shading systems. It is evident that this category mainly pertains to materials and components with selective properties toward light and solar control |
| | | Dynamic façade | Case studies with one or more façades able to vary their performance feature over time. This has been considered as a separate category because previous ones can be prerogative of a single material or component while in this section are intended only façades as a whole |
| | | Adaptive glazing | This category can be considered a specialization of the previous one, although for the specific role that transparent building components play in the present work, it has been separated; within it fit all chromogenic technologies applied to glazed components |
| | | Energy generating | Case studies with particular smart features that allow them to integrate within the envelope energy generation abilities |

(continued)

**Table 4.1** (continued)

| | | | |
|---|---|---|---|
| | | Control system | Category that includes, in a wide sense, all the case studies that allow interaction and users' control |
| | Operation mode | Passive | If the relevant technologies have been activated by means of external stimulus of various nature |
| | | Active | If the relevant technologies have been activated through electrical or mechanical control |
| | Trigger | Thermal | Activation through a temperature gap |
| | | Electromagnetic | Activation due to the presence of an electromagnetic input |
| | | Optical | Activation through a modification in lighting (solar responsive) |
| | | Air quality | Activation through modifications in (generally indoor) air quality, such as, for instance, moisture level or particular values' concentration |
| | | Occupants | Activation through inputs of anthropic nature |
| | | Other | Activation through other kinds of inputs not suitable to other categories identified, such as, for instance, other environmental factors |
| | Response input | Extrinsic | Activation through an external event |
| | | Intrinsic | Automatic activation through environmental stimuli |
| | | Self-adjusting | Self-activation according to external variations |
| | | Users' control | Manual activation by occupants |
| | Degree of adaptability | On/off | If the adaptive feature is not adaptable but just switched on or off |
| | | Gradual | If the adaptive feature is gradual according to different operation scenario |
| | TRL | 1 | Basic principles observed |
| | | 2 | Technology concept formulated |
| | | 3 | Experimental proof of concept |
| | | 4 | Technology validated in lab |
| | | 5 | Technology validated in the relevant environment |
| | | 6 | Technology demonstrated in the relevant environment |
| | | 7 | System prototype demonstration in the operational environment |
| | | 8 | System complete and qualified |
| | | 9 | Actual system proven in the operational environment |

### 4.1.4 *Factsheets*

**Gemini House (1991)**
Roland Mösl, Weiz (A)

**Fig. 4.1** Gemini House prototype built in Weiz as part of the National Energy Exhibition. Details of the independently rotating façade. Source: https://commons.wikimedia.org/wiki/File:Geminihaus_in_Weiz.png by Jonfelat (own work).  via Wikimedia Commons

**Introduction** Conceived by Roland Mösl (a former computer programmer) in 1991 with the purpose to generate solar electricity while performing the common residential function, the Gemini House is a prototype design that rotates to follow the sun, exploiting solar energy to produce electricity.

**The Smart Feature** Constructed to generate more electricity than its inhabitants consume, it was actually the first energy plus house in Austria. The house has a cylindrical plan and rotates to follow the sun; it is extremely insulated and equipped with a heat recovery system; its envelope has up to 150 $m^2$ of photovoltaic panels. The rest of the façade is cladded with ACRYLITE® panels.

**Intended use** Residential | Service industry | Specialized

**Building state of development** Concept/Prototype | Real-scale edifice | Under construction

**Climate** Tropical | Dry/Desertic | Temperate | Continental | Polar

**Application scale** Element/Material | Component | Façade system

**Type** Insulation | Glazing | Shading | Opening systems | Energy harvesting/generation device | Other

**Purpose** Thermal comfort | Visual performance | Acoustic comfort | Energy management/generation | IAQ | Interaction and users' control | Other

**Adaptive function** Collect and storage | Control | Prevent/Reject | Allow interaction

**Classification** High performance | Selective | Dynamic façade | Adaptive glazing | Energy generation | Control system

**Operation mode** Passive | Active

**Trigger** Thermal | Electromagnetic | Optical | Air quality | Occupants | Other

**Response input** Extrinsic | Intrinsic | Self-adjusting | Users' control

**Degree of adaptability** On/Off | Gradual

**TRL** 6

**Heliotrope (1994)**
Rolf Disch, Freiburg (DE)

**Fig. 4.2** Heliotrope, Freiburg. Source: https://commons.wikimedia.org/wiki/File:Heliotrop_Freiburg.jpg by Andrewglaser (own work). Creative Commons Attribution-Share Alike 3.0 Unported license https://creativecommons.org/licenses/by/3.0/deed.en via Wikimedia

**Introduction** First building in the world able to capture more energy than what it consumes, the Heliotrope, an eighteengonal construction, is extensively glazed on one side while opaque isolated on the other. Conceived as a single-axis tracking collector, able to turn according to the sun, its design was replicated by the inventor in other versions of the project that accommodate different utilizations.

**The Smart Feature** The Heliotrope can track the sun, thanks to its building structure that physically rotates to maximize solar heat gains and activates insulation gains to the evacuated tube collector plant during the winter season. In summer, instead, the house turns away from the sun to efficiently prevent overheating; under normal conditions, indeed, the building rotated for 15° per hour. This feature, in combination with the superior insulation of the envelope, allows it to capture approximately an amount around four and six times its energy usage, depending on the time of year.

**The technologies** Its assembly is made of different energy generation modules, including a dual-axis solar photovoltaic tracking panel. All building parts are industrially manufactured with CNC technologies so that the house can be built on-site by connecting the segments to the central column. This manufacturing method allows low-cost serial production of Heliotrope concept.
The transparent components are designed as triple-pane windows with heat-absorbing glass. This significantly reduces heating and cooling requirements for the building throughout the year, which are provided for by a heat pump, while hot water is provided by vacuum-tube solar panels. The walls are well insulated (with a layer of 30 cm), with a U-value in the range of 0.10–0.13.
The photovoltaic solar panels placed on the roof provide five to six times more energy than the building uses, making it energy-positive; to further improve the energy capturing ability, the panels rotate independently from the building to follow the sun and adapt their orientation in case of strong winds.

**The performance** Dynamic simulations show a heating energy demand of only 21 kWh/m$^2$, whereby the active solar systems that reduce the heating energy demand to zero are actually not considered.

**Intended use** Residential | Service industry | Specialized

**Building state of development** Concept/Prototype | Real-scale edifice | Under construction

**Climate** Tropical | Dry/Desertic | Temperate | Continental | Polar

**Application scale** Element/Material | Component | Façade system

**Type** Insulation | Glazing | Shading | Opening systems | Energy harvesting/generation device | Other

**Purpose** Thermal comfort | Visual performance | Acoustic comfort | Energy management/generation | IAQ | Interaction and users' control | Other

**Adaptive function** Collect and storage | Control | Prevent/Reject | Allow interaction

**Classification** High-performance | Selective | Dynamic façade | Adaptive glazing | Energy generation | Control system

**Operation mode** Passive | Active

**Trigger** Thermal | Electromagnetic | Optical | Air quality | Occupants | Other

**Response input** Extrinsic | Intrinsic | Self-adjusting | Users' control

**Degree of adaptability** On/Off | Gradual

**TRL** 8

**Museum of Paper Art (2002)**
Shigeru Ban Architects, Shizuoaka, Japan

**Fig. 4.3** Museum of Paper Art. View of the vertical-shutter façade. Source http://www.shigeruban architects.com/works/2002_paper-art-museum/index.html, photographer Hiroyuki Hirai©, by permission of Shigeru Ban Architecs

**Introduction** The Shigeru Ban's Museum of Paper Art (PAM) is a private museum of a paper manufacturer which exploits contemporary materials (Fiberglass Reinforced Panels) to create modern spaces while maintaining the relationship between the inside and the outside typical of traditional Japanese architecture. The square floor plan is divided into three rows with, in the middle, a three-high story atrium.

**The Smart Feature** The spatial continuity is entrusted by a reinterpretation of the traditional *shitomido*, a vertical window shutter common in traditional Japanese architecture, with opening stacking shutters and awnings. All the external buildings' façades indeed are made of moveable, translucent, and glazed fiber-reinforced panels, assembled on a steel frame. In the east/west sides, the panels are 10-m high while in the south side they can be piloted out for an angle of 90°C.
Panels can be folded, thanks to mechanical guide rails, which, when opened, act as canopies, directing shadow to the interior spaces and their adjacent terraces.

**Intended use** Residential I Service industry I <u>Specialized</u>

**Building state of development** Concept/Prototype I <u>Real-scale edifice</u> I Under construction

**Climate** Tropical I Dry/Desertic I <u>Temperate</u> I Continental I Polar

**Application scale** Element/Material I Component I <u>Façade system</u>

**Type** Insulation I Glazing I Shading I <u>Opening systems</u> I Energy harvesting/generation device I Other

**Purpose** Thermal comfort I Visual performance I Acoustic comfort I Energy management/generation I IAQ I <u>Interaction and users' control</u> I Other

**Adaptive function** Collect and storage I Control I <u>Prevent/Reject</u> I Allow interaction

**Classification** High-performance I Selective I <u>Dynamic façade</u> I Adaptive glazing I Energy generation I Control system

**Operation mode** Passive I <u>Active</u>

**Trigger** Thermal I Electromagnetic I Optical I Air quality I Occupants I <u>Other</u>

**Response input** Extrinsic I Intrinsic I Self-adjusting I <u>Users' control</u>

**Degree of adaptability** On/Off I <u>Gradual</u>

**TRL** 9

**Lumenhaus (2002–2010)**

Virginia Tech team in collaboration with the U.S. Department of Energy, Solar Decathlon Competition 2010, Madrid, Spain.

**Fig. 4.4** The Lumenhaus prototype during the Solar Decathlon competition. (**a**) Source https://www.flickr.com/photos/44085838@N03/4724725840/ by Richard King/U.S. Department of Energy Solar Decathlon. Creative Commons Attribution 2.0 Generic license (https://creativecommons.org/licenses/by/2.0/deed.en). (**b**) Source https://www.flickr.com/photos/solar_decathlon/4006986004/in/album-72157622448785853/ by Dept of Energy Solar Decathlon. CC BY-ND 2.0 (https://creativecommons.org/licenses/by-nd/2.0/) via Flickr

**Introduction** Born as a net-zero home inspired from the Mies Van Der Rohe Farnsworth House, the Lumenhaus was designed by a heterogeneous team built within the Virginia Tech campus to foster collaboration among students of different academic departments. The aim of such a project is to combine the use of alternative technologies, new materials, advances in computer science, and product design, integrating a range of sustainable strategies (compact volume, minimal air infiltration, high insulation level, natural cross-ventilation, and passive heating).
Lumenhaus was exhibited in Washington DC, New York, Chicago, and Madrid, where it won the 2010 Solar Decathlon Europe Competition.

**The Smart Feature** The combination of different technologies resulted in new possibilities for building practice toward modular and flexible architecture. The west façade is realized with an electrochromic glass able to automatically tint to regulate the light within confined spaces; the roof is equipped with a photovoltaic module, while other façades (north and south façades) are equipped with intelligent sliding layers made of insulating panels and metal shutters to filter into the house different kind of lights.
Furthermore, Lumenhaus makes use of an innovative computer monitoring and controls system, using sensors and servomotors linked to centralized computer controls, and the movable shutters and insulating panels are adjusted on an hourly, daily, or seasonal basis to maximize building performance.

**The technologies** This design combines different technologies, with the most relevant certainly the Eclipsis System, made of panels able to automatically open and close, made of two layers: the first is a stainless steel sheet metal, patterned with a geometry of circular laser-cut openings, while the second is a transparent layer made of an innovative wall assembly of two polycarbonate panels, filled with Nanogel insulation. LED panels to lighten the spaces are embedded in the system.

**Intended use** <u>Residential</u> | Service industry | Specialized

**Building state of development** <u>Concept/Prototype</u> | Real-scale edifice | Under construction

**Climate** Tropical | Dry/Desertic | <u>Temperate</u> | Continental | Polar

**Application scale** Element/Material | <u>Component</u> | Façade system

**Type** Insulation | Glazing | <u>Shading</u> | Opening systems | Energy harvesting/generation device | Other

**Purpose** Thermal comfort | <u>Visual performance</u> | Acoustic comfort | Energy management/generation | IAQ | Interaction and users' control | Other

**Adaptive function** Collect and storage | Control | <u>Prevent/Reject</u> | Allow interaction

**Classification** High-performance | Selective | <u>Dynamic façade</u> | Adaptive glazing | Energy generation | Control system

**Operation mode** Passive | <u>Active</u>

**Trigger** Thermal | Electromagnetic | Optical | Air quality | <u>Occupants</u> | Other

**Response input** <u>Extrinsic</u> | Intrinsic | Self-adjusting | Users' control

**Degree of adaptability** <u>On/Off</u> | Gradual

**TRL** 7

**Esplanade Complex (2002)**
DP Architects, James Stirling and Michael Wilford, Singapore

**Fig. 4.5** View of the Theater by the Bay, part of the Esplanade Complex in Singapore. Source: https://commons.wikimedia.org/wiki/File:Singapore_Esplanade_-_Theatres_by_the_Bay_4.jpg by Zairon (own work). Creative Commons Attribution-Share Alike 4.0 International license (https://creativecommons.org/licenses/by-sa/4.0/deed.en) via Wikimedia commons

**Introduction** The Esplanade Complex (built as the center for performing arts) in Singapore is made up of two rounded glass domes fitted with over 7,000 triangular aluminum sunshades, which cover a mainly transparent glass structure.
The particular shapes of the screens preserve the desired architectural expression and views of the complex (which originally provide for a fully glazed design); due to the complex geometry of the volumes, the external layer is made of a module that varies in geometry around the building to allow views outside while still providing the maximum amount of shading.

**The Smart Feature** The triangular shading devices offer an optimized trade-off between solar shading and outside panoramic views. These shading panels were properly designed to filter sunlight but allow an amount of sun to enter the space resorting to computer models that were developed to derive optimized shade geometry for each orientation of the façade. Each leaf had the same base dimension but different height so that the orientation of the surface relative to the sun angle would provide the maximum amount of shade.

The shields, aluminum made, are set to be more open or closed depending on the angle at which the sun hits them, affording the glass façades protection from direct sunlight without limiting the view.

**Intended use** Residential | Service industry | Specialized

**Building state of development** Concept/Prototype | Real-scale edifice | Under construction

**Climate** Tropical | Dry/Desertic | Temperate | Continental | Polar

**Application scale** Element/Material | Component | Façade system

**Type** Insulation | Glazing | Shading | Opening systems | Energy harvesting/generation device | Other

**Purpose** Thermal comfort | Visual performance | Acoustic comfort | Energy management/generation | IAQ | Interaction and users' control | Other

**Adaptive function** Collect and storage | Control | Prevent/Reject | Allow interaction

**Classification** High-performance | Selective | Dynamic façade | Adaptive glazing | Energy generation | Control system

**Operation mode** Passive | Active

**Trigger** Thermal | Electromagnetic | Optical | Air quality | Occupants | Other

**Response input** Extrinsic | Intrinsic | Self-adjusting | Users' control

**Degree of adaptability** On/Off | Gradual

**TRL** 9

**Jubilee Church (2003)**
Richard Meier & Partners, Rome (IT)

**Fig. 4.6** Jubilee Church. Source https://www.flickr.com/photos/wojtekgrak/12742013913/in/photostream/ by Wojtek Gurak. CC BY-NC 2.0 (https://creativecommons.org/licenses/by-nc/2.0/) via Flickr

**Introduction** Conceived as part of Pope John Paul II's millennium initiative, the Jubilee Church is located outside central Rome. Its shape, full of symbolical meanings, is based on a geometrical structure made of squares and circles; the body of the church indeed is generated by the profiles of the three concrete shells that imply the Holy Trinity. Similarly, the reflecting pool located in the front of the complex recalls the role played by water in the sacrament of Baptism, as well as the materials used in the portico, which symbolized the body of Christ's church.

**The Smart Feature** The façades' ashlars are treated with a super brightness photocatalytic concrete, called Millennium TX, specifically developed for this construction by Italcementi, able to maintain its whiteness over time. Experimental tests showed indeed that after about 60 h of sun exposure, samples of such materials—dirty with ash coming from cigarettes—recovered their original color.

**The technologies** The abovementioned effect is obtained through a photocatalytic behavior, based on the use of catalysts able to oxidize harmful substances and pollutants until their complete mineralization. Applied in architecture, these technologies allow to obtain passive air purification by reducing organic and inorganic pollutants, a deodorizing and antimicrobial action, and a self-cleaning action due to the fact that surfaces coated with titanium dioxide are totally water repellent.

**Intended use** Residential | Service industry | Specialized

**Building state of development** Concept/Prototype | Real-scale edifice | Under construction

**Climate** Tropical | Dry/Desertic | Temperate | Continental | Polar

**Application scale** Element/Material | Component | Façade system

**Type** Insulation | Glazing | Shading | Opening systems | Energy harvesting/generation device | Other

**Purpose** Thermal comfort | Visual performance | Acoustic comfort | Energy management/generation | IAQ | Interaction and users' control | Other

**Adaptive function** Collect and storage | Control | Prevent/Reject | Allow interaction

**Classification** High-performance | Selective | Dynamic façade | Adaptive glazing | Energy generation | Control system

**Operation mode** Passive | Active

**Trigger** Thermal | Electromagnetic | Optical | Air quality | Occupants | Other

**Response input** Extrinsic | Intrinsic | Self-adjusting | Users' control

**Degree of adaptability** On/Off | Gradual

**TRL** 9

**EWE Arena (2003)**
wh-p Ingenieure, Oldenburg (DE)

**Fig. 4.7** The EWE Arena. Source: https://commons.wikimedia.org/wiki/File:EWE-Arena.jpg by WilfriedC (own work). CC BY-SA 3.0 (https://creativecommons.org/licenses/by-sa/3.0) via Wikimedia Commons

**Introduction** With around 16,000 $m^2$ of solar cells, the EWE Arena is Germany's largest photovoltaic system integrated into a building. Its exterior walls are totally glazed and, to prevent summer overheating, are shaded by concrete overhangs and by a large movable PV-shading system able to move around the outer façade, according to the sun's position.

**The Smart Feature** The kinetic PV-wall, which provides shading when needed, produces—at the same time—energy by exploiting the photovoltaic effect. The photovoltaic panels indeed can rotate 200° around the outer façade. Their size (6.5 m high and 36 m long) covers a 60° segment of the building, and tracking is done every 30′ in 7.5° steps.

**The technologies** The façade is hanged up on steel rails fixed at the concrete structure. They work as guides and take care of the wind loads. The sunscreen consists of 18 segments, which hold 8 PV modules each; the frameless PV modules are designed to be semitransparent and are built of crystalline cells. The three inverters are mounted directly on the movable sunscreen. The cabling is done with the help of a distributing jumper ring at the top. The whole façade consists of 72 PV Modules and 72 glass elements, with screen printing. The total weight of the movable sunscreen is 28 t.

**Intended use** Residential | Service industry | Specialized

**Building state of development** Concept/Prototype | Real-scale edifice | Under construction

**Climate** Tropical | Dry/Desertic | Temperate | Continental | Polar

**Application scale** Element/Material | Component | Façade system

**Type** Insulation | Glazing | Shading | Opening systems | Energy harvesting/generation device | Other

**Purpose** Thermal comfort | Visual performance | Acoustic comfort | Energy management/generation | IAQ | Interaction and users' control | Other

**Adaptive function** Collect and storage | Control | Prevent/Reject | Allow interaction

**Classification** High-performance | Selective | Dynamic façade | Adaptive glazing | Energy generation | Control system

**Operation mode** Passive | Active

**Trigger** Thermal | Electromagnetic | Optical | Air quality | Occupants | Other

**Response input** Extrinsic | Intrinsic | Self-adjusting | Users' control

**Degree of adaptability** On/Off | Gradual

**TRL** 9

**Offices 'La Defense' (2004)**
UNStudio, Almere, The Netherlands

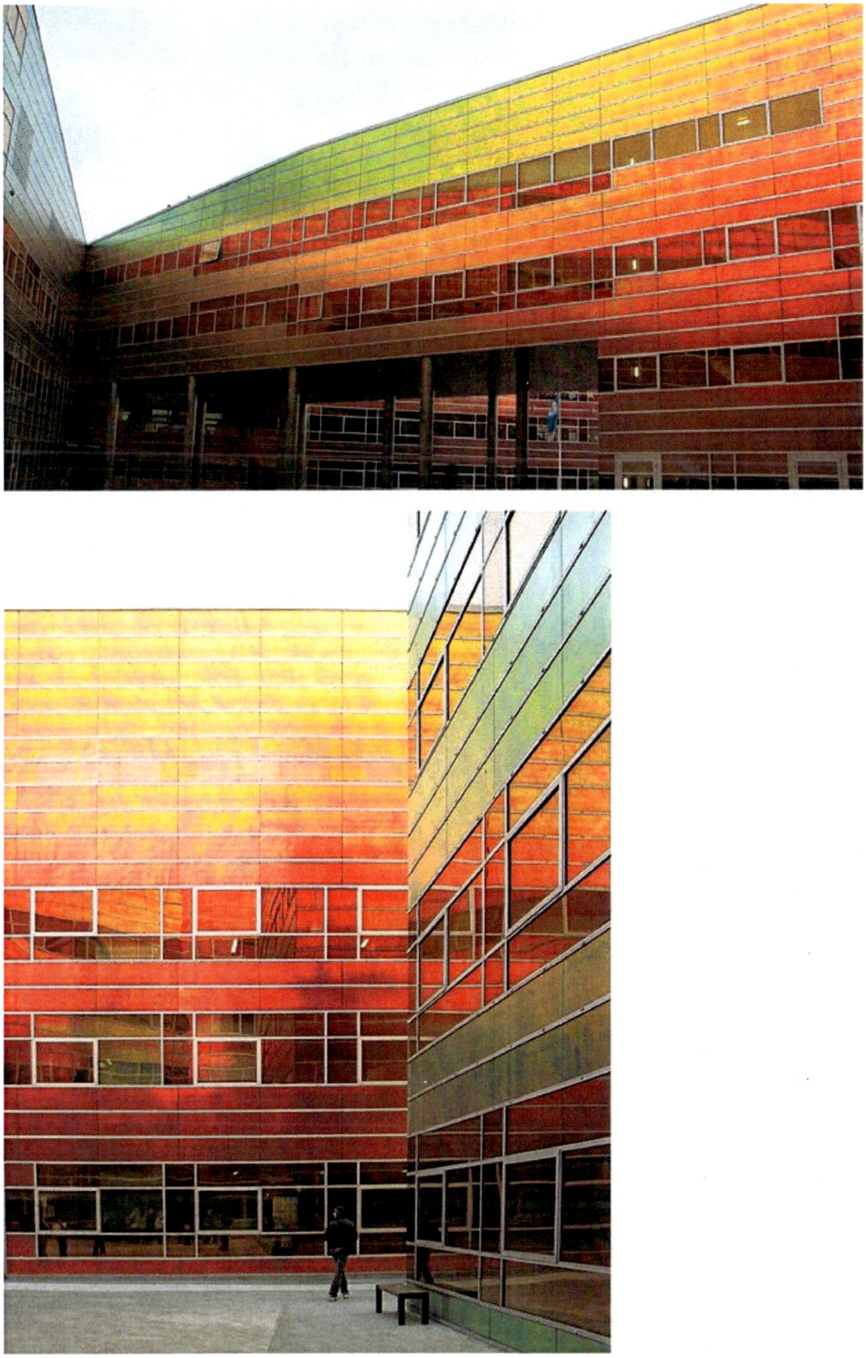

**Fig. 4.8** Particular of the Offices' façades toward the interior courtyard. (**a**) Source https://commons.wikimedia.org/wiki/File:La_Defense_Almere01.jpg by laurenatclemson (own work). CC BY 2.0 (https://creativecommons.org/licenses/by/2.0) via Wikimedia Commons. (**b**) Source https://commons.wikimedia.org/wiki/File:La_Defense_Almere02.jpg by laurenatclemson (own work). CC BY 2.0 (https://creativecommons.org/licenses/by/2.0) via Wikimedia Commons

**Introduction** The project was conceived as a single block of offices, then sectioned leaving some open slots in the center; it is located in a business park surrounded by partially residential plots and behind the central station of the city, uniting them by connecting the texture of the urban fabric. The buildings' façades, indeed, reflect the urban conditions of the neighborhood through a mirroring gray curtain façade, while the interior façades have been properly designed to face offices' well-being requirements, displaying, at the same time, a dichroic skin capable of changing color according to the light sources.

**The Smart Feature** The particular façades' treatment confers to the building the ability to change according to the variability of the external conditions, providing at the same time privacy for its occupants. During the day, indeed, the façade changes from yellow to blue, to red, or from purple to green and back again.

**The technologies** The façades are realized through the use of special glass panels in which a multicolored foil is integrated; depending on the time of the day as well as the sunlight angle of incidence, the foil reflects different colors. This particular film, generally used for packaging, is a multilayered polymeric film with a polyester surface, which confers to the building façade a dichroic effect by affecting light rays: Each boundary between the layers reflects a fraction of the incoming light.

**Intended use** Residential | Service industry | Specialized

**Building state of development** Concept/Prototype | Real-scale edifice | Under construction

**Climate** Tropical | Dry/Desertic | Temperate | Continental | Polar

**Application scale** Element/Material | Component | Façade system

**Type** Insulation | Glazing | Shading | Opening systems | Energy harvesting/generation device | Other

**Purpose** Thermal comfort | Visual performance | Acoustic comfort | Energy management/generation | IAQ | Interaction and users' control | Other

**Adaptive function** Collect and storage | Control | Prevent/Reject | Allow interaction

**Classification** High-performance | Selective | Dynamic façade | Adaptive glazing | Energy generation | Control system

**Operation mode** Passive | Active

**Trigger** Thermal | Electromagnetic | Optical | Air quality | Occupants | Other

**Response input** Extrinsic | Intrinsic | Self-adjusting | Users' control

**Degree of adaptability** On/Off | Gradual

**TRL** 9

**Chanel Ginza (2004)**
Peter Marino Architect, Tokyo (JP)

**Fig. 4.9** Transition of the Chanel Ginza façade between day and night. Source: https://commons.wikimedia.org/wiki/File:Chanel-Ginza.jpg by Luc Boegly COM:OTRS. CC BY-SA 4.0 (https://creativecommons.org/licenses/by-sa/4.0) via Wikimedia Commons

**Introduction** This 10-story high media façade couples views and daylight priorities with the communication needs of the design house. It was the first media building developed with an integrated LED system inside a glass curtain wall.

**The Smart Feature** The Chanel Ginza building resolves both the concepts of solar control and the many competing functions of the illuminated walls. During the day, the façade is transparent and provides views to the interior of the building (since the internal blinds are retracted); during the night, instead, the blinds provide a back background for the video images, and the system transforms the façade into a giant projection screen, thanks to hundreds of thousands of LEDs sandwiched between a triple layer of glass.

**The technologies** The façade is primarily made of gray-tinted Low-E laminated glass, which provides for solar control, that incorporates—within the triple-glazed wall units—a stainless steel mesh. The louvers of the metal layer provide the desired textile effects, offering at the same time another layer of solar shading. Moreover, the metal mesh serves to control light spill from each pixel of the white LED array, placed outside a motorized canvas roller shade. Tubes spanning small vertical supports contain four rows of LEDs, with a mix of narrow and wide beam spreads; two are aimed upward and two point downward, and the combined light of 72 LEDs constitutes a single 8-inch-wide pixel. The main façade is 188 pixels tall by 98 pixels wide.

**Intended use** Residential I Service industry I Specialized

**Building state of development** Concept/Prototype I Real-scale edifice I Under construction

**Climate** Tropical I Dry/Desertic I Temperate I Continental I Polar

**Application scale** Element/Material I Component I Façade system

**Type** Insulation I Glazing I Shading I Opening systems I Energy harvesting/generation device I Other

**Purpose** Thermal comfort I Visual performance I Acoustic comfort I Energy management/generation I IAQ I Interaction and users' control I Other

**Adaptive function** Collect and storage I Control I Prevent/Reject I Allow interaction

**Classification** High-performance I Selective I Dynamic façade I Adaptive glazing I Energy generation I Control system

**Operation mode** Passive I Active

**Trigger** Thermal I Electromagnetic I Optical I Air quality I Occupants I Other

**Response input** Extrinsic I Intrinsic I Self-adjusting I Users' control

**Degree of adaptability** On/Off I Gradual

**TRL** 8

**Selimex Offices (2005)**
Werner Tscholl, Laces, Bozen (IT)

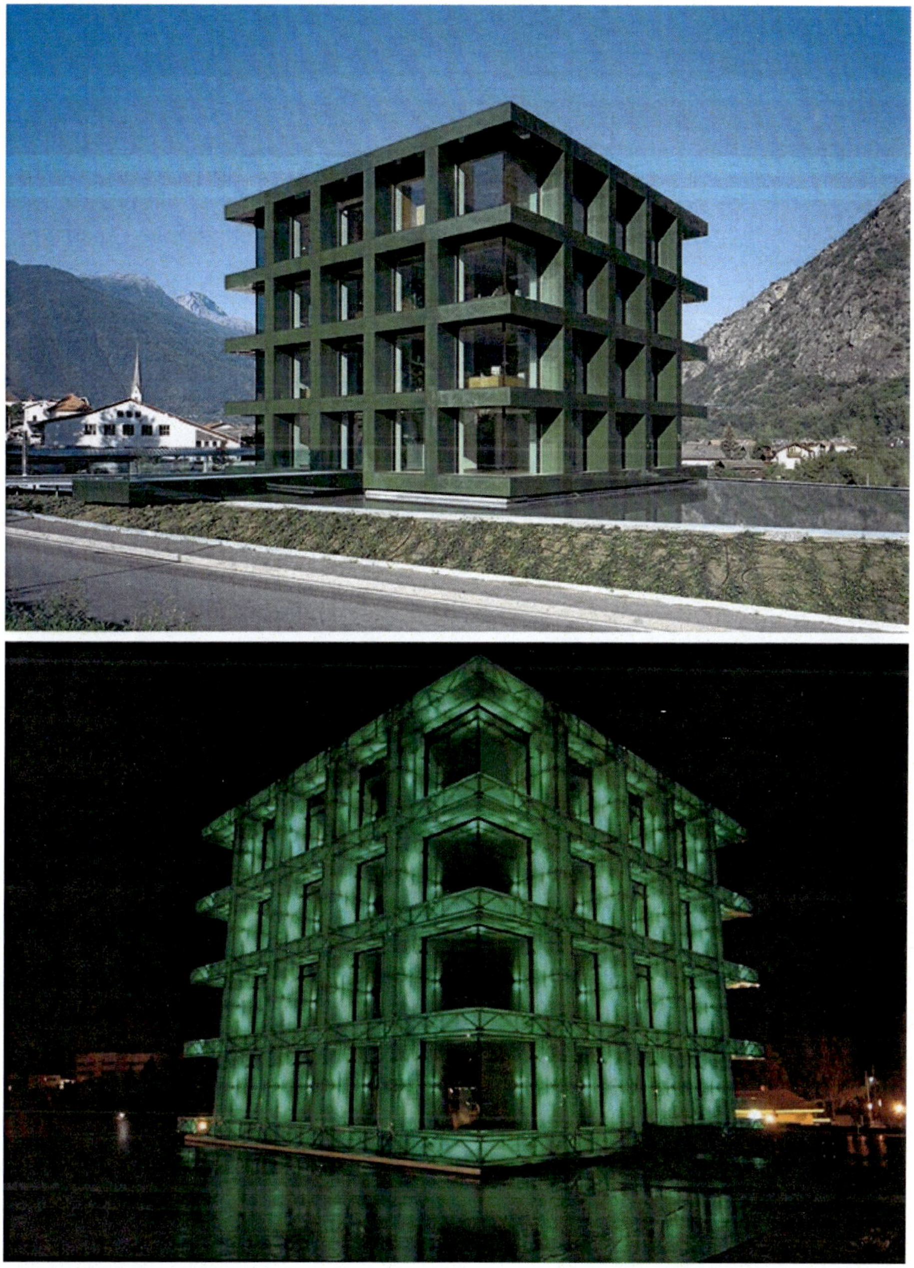

**Fig. 4.10** Selimex Offices during the day (**a**, above) and night (**b**, below). (Courtesy of Werner Tscholl)

**Introduction** Located right at the entrance of the Latsch-Laces village, the Selimex head offices represent a significant mark in the surrounding, both for its composition and for its materials. Also, the presence of water emphasizes the role of the building, which seems arising from it. This is also a symbol of the firm that works in it, which deals with import–export, as it may be seen like a ship laying alongside a harbor.
The completely glazed parallelepiped is characterized on the sides by volume protruding horizontally and vertically, glazed as well.

**The Smart Feature** Completely made of glass, during the day the proximity of water (so as the reflections of surrounding surfaces) lets the building to match perfectly with the site, and during the night, the artificial lights give it continuously changing aspects.
To ensure this feature, special spotlights have been created with high-power LED technology to yield almost endless color combinations; furthermore, they can be programmed to permit gradually but continuously changing chromatic series.

**Intended use** Residential | Service industry | Specialized

**Building state of development** Concept/Prototype | Real-scale edifice | Under construction

**Climate** Tropical | Dry/Desertic | Temperate | Continental | Polar

**Application scale** Element/Material | Component | Façade system

**Type** Insulation | Glazing | Shading | Opening systems | Energy harvesting/generation device | Other

**Purpose** Thermal comfort | Visual performance | Acoustic comfort | Energy management/generation | IAQ | Interaction and users' control | Other

**Adaptive function** Collect and storage | Control | Prevent/Reject | Allow interaction

**Classification** High-performance | Selective | Dynamic façade | Adaptive glazing | Energy generation | Control system

**Operation mode** Passive | Active

**Trigger** Thermal | Electromagnetic | Optical | Air quality | Occupants | Other

**Response input** Extrinsic | Intrinsic | Self-adjusting | Users' control

**Degree of adaptability** On/Off | Gradual

**TRL** 9

**Kiefer Technic Showroom (2007)**
Ernst Giselbrecht + Partner, Bad Gleichenberg, Austria

**Fig. 4.11** Ernst Giselbrecht + Partner (architects), Detail of the Kiefer Technic façade in action. Photographer Paul Ott© by permission of Ernst Giselbrecht + Partner. For further details about the façade mode of operation see the video "Eine Fassade tanzt" on YouTube (https://youtu.be/rAn4ldWjw2w)

**Introduction** The showroom, an office building in Bad Gleichenberg, presents the first example of a user-controlled dynamic façade shaped with folding elements made of perforated aluminum electrically driven.

**The Smart Feature** The dynamic façade is able to change according to outdoor conditions and to optimize internal climate while allowing users to personalize indoor spaces. The white panels, indeed, are able to expand or contract vertically (they can fold into a horizontal shading position), regulating the radiance of sunlight. The shading devices can be controlled in a manual way but also by optimized programs if users are not present.

**The technologies** The opaque buildings' façades are made of solid brick walls, while the transparent ones present great striped windows shaded with electronic shutters made of 112 preformed aluminum panels. Each panel is made of a perforated, powder-coated light metal cassette that can be moved in all three dimensions. Such dynamic feature is ensured by the presence of electronic controls that can control each of the 54 motors within the façade. The panels can also be individually controllable. There are no responsive systems or responses because the only inputs are those given from the building occupants; the façade indeed is functioning as a shading device whose angle is decided by users, which establishes the amount of light transmitted into the interior spaces.

**Intended use** Residential I Service industry I Specialized

**Building state of development** Concept/Prototype I Real-scale edifice I Under construction

**Climate** Tropical I Dry/Desertic I Temperate I Continental I Polar

**Application scale** Element/Material I Component I Façade system

**Type** Insulation I Glazing I Shading I Opening systems I Energy harvesting/generation device I Other

**Purpose** Thermal comfort I Visual performance I Acoustic comfort I Energy management/generation I IAQ I Interaction and users' control I Other

**Adaptive function** Collect and storage I Control I Prevent/Reject I Allow interaction

**Classification** High-performance I Selective I Dynamic façade I Adaptive glazing I Energy generation I Control system

**Operation mode** Passive I Active

**Trigger** Thermal I Electromagnetic I Optical I Air quality I Occupants I Other

**Response input** Extrinsic I Intrinsic I Self-adjusting I Users' control

**Degree of adaptability** On/Off I Gradual

**TRL** 9

**Pola Ginza building (2009)**
Nikken Sekkei + Yasuda Atelier, Ginza district, Tokyo, Japan

**Fig. 4.12** View of the Pola building from the main street. Source: https://commons.wikimedia.org/wiki/File:POLA_Ginza_Buildings.JPG by ITA-ATU (own work). CC BY-SA 4.0 (https://creativecommons.org/licenses/by-sa/4.0) via Wikimedia Commons

**Introduction** Mixed-use tower in Tokyo's Ginza (it houses a showroom for cosmetics company POLA on the lower floors and commercial office space on the upper levels) with a double-skin glazed façade with kinetic panels, designed in collaboration with the Adaptive Building Initiatives and Hobermann Associates.

**The Smart Feature** The adaptive shading system, composed of about 185 elements—operable to shield the interiors from direct sunlight—presents the ability to change size and shape.

**The technologies** Each shutter, sized approximately 1 × 3 m, is composed of a double-glazed system that embeds a kinetic polycarbonate layer that renders it translucent; the acrylic sheet has been formed into a curved surface and presents different patterns that illustrate the cells of life.
The shutters operate in groups of 40 on different mechanisms hasted within the double glazing of the façade. A building management system coupled with photo sensors controls the translucent shading devices, helping in reducing heat gain during the seasons. The double glazing system reduces annual HVAC energy by 30%.

**Intended use** Residential I Service industry I Specialized

**Building state of development** Concept/Prototype I Real-scale edifice I Under construction

**Climate** Tropical I Dry/Desertic I Temperate I Continental I Polar

**Application scale** Element/Material I Component I Façade system

**Type** Insulation I Glazing I Shading I Opening systems I Energy harvesting/generation device I Other

**Purpose** Thermal comfort I Visual performance I Acoustic comfort I Energy management/generation I IAQ I Interaction and users' control I Other

**Adaptive function** Collect and storage I Control I Prevent/Reject I Allow interaction

**Classification** High-performance I Selective I Dynamic façade I Adaptive glazing I Energy generation I Control system

**Operation mode** Passive I Active

**Trigger** Thermal I Electromagnetic I Optical I Air quality I Occupants I Other

**Response input** Extrinsic I Intrinsic I Self-adjusting I Users' control

**Degree of adaptability** On/Off I Gradual

**TRL** 8

**North House (2009)**
Team North (University of Waterloo, Ryerson University and Simon Fraser University), US Department of Energy Solar Decathlon, Washington DC (USA).

**Fig. 4.13** The North House during the Solar Decathlon competition. Source: https://www.flickr.com/photos/solar_decathlon/4006224031/in/album-72157622448785853/ by Dept, of Energy Solar Decathlon via Flickr. CC BY-ND 2.0 (https://creativecommons.org/licenses/by-nd/2.0/)

**Introduction** Interdisciplinary research project aimed at developing a full-scale prototype of a net positive energy residence, prefabricated, and optimized for cold climate applications. The goal of this experimentation is to pair hybrid integrated active and passive envelope systems with interactive controls, delivering a net energy-producing model able to foster new relationships among occupants, the building system, and the environment.

**The Smart Feature** Equipped with energy generation capacity, the North House produces annually more energy than what it consumes, thanks to BIPVs and BAPVs (Building Applied Photovoltaics). Its distinctive features are the assemblies developed by the research team, called DReSS (Distributed Responsive System of Skins), CHAS (Central Home Automation Server), and ALIS (Adaptive Living Interface System). The components of DReSS include two-stage operable exterior Venetian shading, high-performance glazing, interior blinds, and floor-integrated phase-change materials along with building-integrated photovoltaic on the exterior cladding. The control system manages the shading to prioritize passive thermal management based on sensor data; the exterior shade configuration scenarios are based on relative environmental conditions and related responsive envelope reactions that formed the framework for the development of the home automation system. Besides, CHAS manages energy production and grid connection with the mechanical demands of the North House HVAC system and responsive envelope, while ALIS is basically a BMS that simplifies occupants' use of complex HVAC, energy production, and responsive envelope components. The system combines a control interface with feedback and allows remote operation via smartphone technology. Thanks to these devices, changes can be made and monitored in real time. Energy usage is also minimized by the smart interior design, which incorporates elements such as a custom cellular ceiling, made from window-shade material, that helps to absorb reflected sound and direct light from the perimeter into the core of the space.

**Intended use** Residential I Service industry I Specialized

**Building state of development** Concept/Prototype I Real-scale edifice I Under construction

**Climate** Tropical I Dry/Desertic I Temperate I Continental I Polar

**Application scale** Element/Material I Component I Façade system

**Type** Insulation I Glazing I Shading I Opening systems I Energy harvesting/generation device I Other

**Purpose** Thermal comfort I Visual performance I Acoustic comfort I Energy management/generation I IAQ I Interaction and users' control I Other

**Adaptive function** Collect and storage I Control I Prevent/Reject I Allow interaction

**Classification** High-performance I Selective I Dynamic façade I Adaptive glazing I Energy generation I Control system

**Operation mode** Passive I Active

**Trigger** Thermal I Electromagnetic I Optical I Air quality I Occupants I Other

**Response input** Extrinsic I Intrinsic I Self-adjusting I Users' control

**Degree of adaptability** On/Off I Gradual

**TRL** 3

**Surry Hills Library and Community Centre (2009)**
FJMT, Sydney (AUS)

**Fig. 4.14** Surry Hills Library's façade detail. Source: https://commons.wikimedia.org/wiki/File:Surry_Hills_Library_2010.jpg by Elekhh (own work). CC BY-SA 3.0 (https://creativecommons.org/licenses/by-sa/3.0) via Wikimedia Commons

**Introduction** Designed to achieve excellence in sustainable design and set new high standards in environmental performance for multipurpose public buildings, the building, which won awards for its environment-conscious design, includes a library, a child care center, community rooms, and an art and crafts space for almost 2500 $m^2$. To achieve the ambitious sustainability goal of the project, it includes a number of energy-efficient strategies becoming an innovative example of sustainable design possibilities.

**The Smart Feature** The main relevant feature of the project is the internal glazed atrium and the differing feature façades that tackle building's surrounding each with a different approach. The glazed atrium, surrounded by glass prisms, favors the displaying and transparency of the different activities of the center, encouraging participation. Moreover, it helps to passively heat and cool the center by serving as a buffer to the outside, allowing, at the same time, to exploit natural filtering properties of vegetation as part of a thermal labyrinth for passive filtering and tempering of the air. The wooden façade louvers, sourced from sustainable forests, create a variable-changing façade able to reduce energy cost, thanks to its ability to allow the control of natural light flow in appropriate situations. The shading devices indeed are automatically controlled to move and track the sun.

The environmental initiatives intrinsic to the design also include automated fabric shading devices, mixed-mode ventilation, extensive photovoltaic array, geothermal cooling bores, green roof, rainwater collection and recycling, and, of course, a sustainable material selection. Computerized building management and control systems (BMS) automatically monitor and control the internal environmental conditions of the building, adjusting the ventilation and sunshade louvers throughout the day to control heat load, light and shade, and switching lights on and off when required. The BMS also monitors and records both electrical and hydraulic systems to maximize the environmental efficiency of the building and identify system faults.

**Intended use** Residential | Service industry | Specialized

**Building state of development** Concept/Prototype | Real-scale edifice | Under construction

**Climate** Tropical | Dry/Desertic | Temperate | Continental | Polar

**Application scale** Element/Material | Component | Façade system

**Type** Insulation | Glazing | Shading | Opening systems | Energy harvesting/generation device | Other

**Purpose** Thermal comfort | Visual performance | Acoustic comfort | Energy management/generation | IAQ | Interaction and users' control | Other

**Adaptive function** Collect and storage | Control | Prevent/Reject | Allow interaction

**Classification** High-performance | Selective | Dynamic façade | Adaptive glazing | Energy generation | Control system

**Operation mode** Passive | Active

**Trigger** Thermal | Electromagnetic | Optical | Air quality | Occupants | Other

**Response input** Extrinsic | Intrinsic | Self-adjusting | Users' control

**Degree of adaptability** On/Off | Gradual

**TRL** 9

**100 11th Avenue (Sunset façade) (2010)**
Atelier Jean Nouvel, New York City, USA.

**Fig. 4.15** The sunset façade from the 11th Avenue. Source: https://commons.wikimedia.org/wiki/File:Jean_Nouvel%27s_100_11th_Avenue.jpg) by DFranck (own work). CC BY-SA 3.0 (https://creativecommons.org/licenses/by-sa/3.0) or GNU Free Documentation License via Wikimedia commons

**Introduction** With a number of approximately 1650 different windowpanes, the Nouvel's building can be considered the most highly engineered and complex curtain wall ever constructed in New York City. His approach aimed at emphasized windows' mullions to render them scattered frames; furthermore, to simplify the construction process, the façade has a fragmented surface with various window dimensions that allow the curving of the curtain wall, each of them is set in a unique angle and torque.

**The Smart Feature** The project has the ability to overcome common curtain walls with a simplified solution derived from the fragmentation of the façade; in addition, different coatings added to glass allow light to play with the façade, continuously changing its appearance. The ability to change brightness and darkness is ensured by sunlight variations; dynamic movements are here ensured by earth movements.

**The technologies** The project employs over 1650 different E-glass panels, curved with a different tilt angle across its surface; the windows are both fixed and operable, in various sizes and shapes; to give regularly to such a system, the manufactures crafted special mega panels that contain smaller, individual panels.

**Intended use** Residential | Service industry | Specialized

**Building state of development** Concept/Prototype | Real-scale edifice | Under construction

**Climate** Tropical | Dry/Desertic | Temperate | Continental | Polar

**Application scale** Element/Material | Component | Façade system

**Type** Insulation | Glazing | Shading | Opening systems | Energy harvesting/generation device | Other

**Purpose** Thermal comfort | Visual performance | Acoustic comfort | Energy management/generation | IAQ | Interaction and users' control | Other

**Adaptive function** Collect and storage | Control | Prevent/Reject | Allow interaction

**Classification** High-performance | Selective | Dynamic façade | Adaptive glazing | Energy generation | Control system

**Operation mode** Passive | Active

**Trigger** Thermal | Electromagnetic | Optical | Air quality | Occupants | Other

**Response input** Extrinsic | Intrinsic | Self-adjusting | Users' control

**Degree of adaptability** On/Off | Gradual

**TRL** 9

**GEOX Megastore (The breath building) (2010)**
Dante Benini & Partners | Architects, Milan (IT)

**Fig. 4.16** View from the street (**a**) and detail of the "breathing" façade of the GEOX Megastore (**b**). Architectural project: Dante O. Benini & Partners | Architects. Project Architects: Dante O. Benini, Luca Gonzo. Photographer Beppe Raso© (Courtesy of Dante O. Benini & Partners | Architects)

**Introduction** Refurbishment project in which only the building envelope has been substituted while the bearing concrete structure has been preserved. The new façade is made of over 100 metal sheets, autumn-colored, and able to open and close, offering a dynamic scenery from the street; this effect increases from a maxi screen that through LED technologies reflect different graphics.

**The Smart Feature** The metal panels not only have an aesthetic value but are rather the second skin that protects the internal environment from overheating or cooling, creating an air gap between the external cladding and the internal walls. In addition, the single-skin walls are coated with a photocatalytic plaster, with chemical properties that transform air pollutants into dust, thus leaving the façade more clean than usual.

**The technologies** The new façade has been built with dry technologies able to support both fenestration and walls; the process so industrialized has allowed lower times and cost when compared to traditional solutions. The microperforated metal sheets are electrically controlled through a device that allows users to interact with the shadings, choosing its angle depending on the incidence of sunlight.

**Intended use** Residential | Service industry | Specialized

**Building state of development** Concept/Prototype | Real-scale edifice | Under construction

**Climate** Tropical | Dry/Desertic | Temperate | Continental | Polar

**Application scale** Element/Material | Component | Façade system

**Type** Insulation | Glazing | Shading | Opening systems | Energy harvesting/generation device | Other

**Purpose** Thermal comfort | Visual performance | Acoustic comfort | Energy management/generation | IAQ | Interaction and users' control | Other

**Adaptive function** Collect and storage | Control | Prevent/Reject | Allow interaction

**Classification** High-performance | Selective | Dynamic façade | Adaptive glazing | Energy generation | Control system

**Operation mode** Passive | Active

**Trigger** Thermal | Electromagnetic | Optical | Air quality | Occupants | Other

**Response input** Extrinsic | Intrinsic | Self-adjusting | Users' control

**Degree of adaptability** On/Off | Gradual

**TRL** 9

**ThyssenKrupp Corporate Headquarter (2010)**
JSWD Architekten, Chaix & Morel et Associès, Essen (DE)

**Fig. 4.17** View of the main façade. Source: https://commons.wikimedia.org/wiki/File:Thyssen-Krupp-Quartier-Essen-Q1-2013.jpg by Tuxyso (own work). CC BY-SA 3.0 (https://creativecommons.org/licenses/by-sa/3.0) via Wikimedia Commons

**Introduction** Born to flexibly react to changes within the company, the new ThyssenKrupp Headquarters is composed of different buildings; the Q1, the heart of the campus, is the most representative, thanks to its shape and to its particularly sustainable features.

The particular structure of the complex, with trees, paths, and squares, makes it able to react flexibly to changes within the corporation.

**The Smart Feature** The smart feature of this project is its ability to avoid air conditioning systems, thanks to the particular glazed structures supported by cable constructions. The great idea was to combine a constant horizontal overhang (useful for summer protection and as a catwalk) with a vertical set of twisting lamellas, able to achieve an adjustable position between 0° and 90°. The shading device, made of about 400,000 stainless steel louvers and fins with triangular, square, and trapezoidal shapes, ensures the solar control of incident radiation, thanks to lamellas' orientation as well as their movements, based on the sun's path to maximize views out (thanks to automatic adjustments of the system) while reducing glare and cutting down on heat gain. A weather station on the roof, indeed, sends signals to a computer that steers the rotation of the façade slats. The programming is really sensitive: The control system not only detects the seasonal sun position but also knows what the current weather is like due to the data of a weather station located on the roof of the building.
Furthermore, the company, to replace expensive aluminum profiles, has developed the Nirosta material, employed in the complex design, a steel sheet with a zinc and magnesium coating.
Another interesting touch was the shape of the fins, which was nonrectangular, thus creating an interesting texture as the element rotates along the day.

**Intended use** Residential | <u>Service industry</u> | Specialized

**Building state of development** Concept/Prototype | <u>Real-scale edifice</u> | Under construction

**Climate** Tropical | Dry/Desertic | <u>Temperate</u> | Continental | Polar

**Application scale** Element/Material | <u>Component</u> | Façade system

**Type** Insulation | Glazing | <u>Shading</u> | Opening systems | Energy harvesting/generation device | Other

**Purpose** Thermal comfort | <u>Visual performance</u> | Acoustic comfort | Energy management/generation | IAQ | Interaction and users' control | Other

**Adaptive function** Collect and storage | <u>Control</u> | Prevent/Reject | Allow interaction

**Classification** High-performance | <u>Selective</u> | Dynamic façade | Adaptive glazing | Energy generation | Control system

**Operation mode** <u>Passive</u> | Active

**Trigger** Thermal | Electromagnetic | <u>Optical</u> | Air quality | Occupants | Other

**Response input** Extrinsic | <u>Intrinsic</u> | Self-adjusting | Users' control

**Degree of adaptability** On/Off | <u>Gradual</u>

**TRL** 9

**Adaptive Solar Façade (2011)**
Team leaded by Prof. Dr. Arno Schlueter, ETH Zurich (CH)

**Fig. 4.18** Images of movable/adaptable PV modules in front of the glass façade, Empa, Dübendorf. (**a**) Photographer Marco Carocari© (**b**) Photographer Arno Schlueter©. Courtesy of Arno Schlueter

**Introduction** Prototype developed by the researchers of the ETH that combines recent developments in architecture, energy technology, and robotics to retain a comfortable indoor climate within confined spaces over year. Actually, it is one of the largest objects showcasing soft robotics in architecture. Clearly, its efficiency depends on the relationships between the façade and the other building components. The Adaptive Solar Façade consists of individual modules mounted on a cable network on the façade; by adjusting the amount of solar radiation through the window, they can control the light and heat of the internal space, and solar energy generation can be optimized.

**The Smart Feature** The smart feature of this case study stands in the multifunctionality of the building envelope coupled with its ability to provide a dynamic appearance, thanks to the modules that made it up, continuously reacting to the external environment. Therefore, its extremely lightness makes this technology suitable even on existing buildings.

**The technologies** Its assembly is made of movable photovoltaic modules mounted on a lightweight structure; their function is twofold because they do not only produce energy while they offer shading and daylight control. The façade is equipped even to store energy and to transform it into pressurized air to be further used to adjust the movable modules when no energy is generated (as, for instance, in case of bad weather). The modules are moveable, thanks to a soft robotic actuator (generally used in bio-mimetic robots) made of a flexible material that changes its form when the pressure in a special chamber changes. With this particular technology, each actuator can be controlled individually and rotated on two axes; this enables the modules to track the sun's movements and to maximize power generation. In addition, an intelligent controller allows the façade to adapt to changing weather conditions and the habits and wishes of the user.

**The Performance** The smart feature of this case study stands in the multifunctionality of the building envelope coupled with its ability to provide a dynamic appearance, thanks to the modules that made it up, continuously reacting to the external environment. Therefore, its extreme lightness makes this technology suitable even on existing buildings.

**Intended use** Residential | Service industry | Specialized

**Building state of development** Concept/Prototype | Real-scale edifice | Under construction

**Climate** Tropical | Dry/Desertic | Temperate | Continental | Polar

**Application scale** Element/Material | Component | Façade system

**Type** Insulation | Glazing | Shading | Opening systems | Energy harvesting/generation device | Other

**Purpose** Thermal comfort | Visual performance | Acoustic comfort | Energy management/generation | IAQ | Interaction and users' control | Other

**Adaptive function** Collect and storage | Control | Prevent/Reject | Allow interaction

**Classification** High-performance | Selective | Dynamic façade | Adaptive glazing | Energy generation | Control system

**Operation mode** Passive | Active

**Trigger** Thermal | Electromagnetic | Optical | Air quality | Occupants | Other

**Response input** Extrinsic | Intrinsic | Self-adjusting | Users' control

**Degree of adaptability** On/Off | Gradual

**TRL** 6

**Media-ICT Building (2011)**
CLOUD 9 – Enric Ruiz-Geli, Barcelona (SP)

**Fig. 4.19** Main façade of the Media-ICT Buildings. Source: https://commons.wikimedia.org/wiki/File:35_Edifici_Media-TIC,_c._Sancho_de_Ávila.JPG by Enfo (own work). CC BY-SA 3.0 (https://creativecommons.org/licenses/by-sa/3.0) via Wikimedia Commons

**Introduction** Commissioned by the Consortium of Zona Franca CZFB and 22@ (an experimental district with a powerful, distributed and accessible, energy load) of Barcelona, the project is entirely based on ICT with an aim to totally sync with users' preference. The construction was built from the top to become transparent at the bottom, thus reducing its impact on the street.

**The Smart Feature** The Media-ICT is designed to be a generator of energy and to optimize its use. The façade cladding, made of ETFE, performs simultaneously as external protection and a mobile solar screen to regulate both light and temperature, thus contributing to decrease building energy demand.

**The technologies** The south-east façade is made of inflatable ETFE cushions that act as sunscreen, opening in winter and closing in summer; they are constructed with three layers of materials, with constant pressure and variable circulation of air between the chambers: The first layer is transparent, while the second and the third have reverse pattern design which, when inflated and joined together, become opaque. The system activates itself automatically according to a temperature sensor network. Each cushion has its own sensors that are controlled separately. Another façade, the south-west facing, is based on a lenticular configuration in which, within an ETFE cushion (actually into a number of ETFE longitudinal bags), a nitrogen-based fog is introduced to obtain a greater opacity able to preserve the privacy of occupants while reducing solar heat gain through the envelope. This technology in fact has allowed us to bring the solar factor down from 0.45 to 0.10. Even on this side, both the envelope and the internal spaces are equipped with environmental sensors that collect information to adjust interior conditions.

**Intended use** Residential | <u>Service industry</u> | Specialized

**Building state of development** <u>Concept/Prototype</u> | Real-scale edifice | Under construction

**Climate** Tropical | Dry/Desertic | <u>Temperate</u> | Continental | Polar

**Application scale** Element/Material | Component | <u>Façade system</u>

**Type** Insulation | Glazing | Shading | Opening systems | Energy harvesting/generation device | <u>Other</u>

**Purpose** <u>Thermal comfort</u> | Visual performance | Acoustic comfort | Energy management/generation | IAQ | Interaction and users' control | Other

**Adaptive function** <u>Collect and storage</u> | Control | Prevent/Reject | Allow interaction

**Classification** High-performance | Selective | Dynamic façade | Adaptive glazing | Energy generation | <u>Control system</u>

**Operation mode** <u>Passive</u> | Active

**Trigger** Thermal | Electromagnetic | <u>Optical</u> | Air quality | Occupants | Other

**Response input** Extrinsic | Intrinsic | Self-adjusting | <u>Users' control</u>

**Degree of adaptability** On/Off | <u>Gradual</u>

**TRL** 7

## F24 Commercial building (2011)
PETERSENARCHITEKTEN, Berlin (DE)

**Fig. 4.20** F40 Commercial building façade on the main street. Source: https://commons.wikimedia.org/wiki/File:F40_3728.jpg by DedicatedToArchitecture (own work). CC BY-SA 4.0 (https://creativecommons.org/licenses/by-sa/4.0) via Wikimedia Commons

**Introduction** Built in place of a two-story house, the Friedrichstrasse 40 Office Building stands within an area rich in culture, history, and art.

**The Smart Feature** The two façades of the building are treated in significant ways. The inner courtyard face has operable shade screens that cover the glazed façade toward the morning glare. Roll-out textile shades can be extended when needed or retracted for more light once the sun has moved past its zenith. The street-facing façade features three 2-story-high windows, perfectly transparent, with a specially designed double-glazed façade made of vertical glass lamellas to protect against heating.

**Intended use** Residential I <u>Service industry</u> I Specialized

**Building state of development** Concept/Prototype I <u>Real-scale edifice</u> I Under construction

**Climate** Tropical I Dry/Desertic I Temperate I <u>Continental</u> I Polar

**Application scale** Element/Material I <u>Component</u> I Façade system

**Type** Insulation I Glazing I <u>Shading</u> I Opening systems I Energy harvesting/generation device I Other

**Purpose** Thermal comfort I <u>Visual performance</u> I Acoustic comfort I Energy management/generation I IAQ I Interaction and users' control I Other

**Adaptive function** Collect and storage I Control I <u>Prevent/Reject</u> I Allow interaction

**Classification** High-performance I <u>Selective</u> I Dynamic façade I Adaptive glazing I Energy generation I Control system

**Operation mode** <u>Passive</u> I Active

**Trigger** Thermal I Electromagnetic I Optical I Air quality I Occupants I <u>Other</u>

**Response input** Extrinsic I <u>Intrinsic</u> I Self-adjusting I Users' control

**Degree of adaptability** <u>On/Off</u> I Gradual

**TRL** 9

**Quadracci Pavilion, Milwaukee Art Museum (2011)**
Santiago Calatrava, Milwaukee (USA)

**Fig. 4.21** Details of the Quadracci Pavilion at Milwaukee Art Museum. Source: https://commons.wikimedia.org/wiki/File:Milwaukee_Art_Museum_1_(Mulad).jpg by Michael Hicks (Mulad) (own work). CC BY 2.0 (https://creativecommons.org/licenses/by/2.0) via Wikimedia Commons

**Introduction** Part of the Milwaukee Art Museum (the largest art museum in the state of Wisconsin), the Burke Brise Soleil, is a moving wing structure featured on top of the Quadracci Pavillon, a sculptural addition to the museum itself, called the Windhover Hall.

**The Smart Feature** The Burke Brise Soleil is a moveable sunscreen with a 66-m wingspan made of 72 steel fins, ranging in length from approximately 8–32 m. As for many other Smart Architecture, the architect takes inspiration from nature in design: Mobile blinds indeed are inspired by the wings of a bird, while the other architectural elements (the bridge and the curved gallery) recall maritime elements (respectively, a sailboat and a wave). The fins have constant cross section width of 330 mm but varying in length, depth, and thickness.
Sensors on the fins continually monitor wind speed and direction, so when the winds exceed 23 mph for more than 3 s, the wings close automatically, controlling in this way the level of air and light to the interior of the museum while creating a sort of kinetic urban sculpture.

The entire structure weighs 90 tons. For the entire span to either open or close, it takes roughly three and a half minutes.

**Intended use** Residential | Service industry | Specialized

**Building state of development** Concept/Prototype | Real-scale edifice | Under construction

**Climate** Tropical | Dry/Desertic | Temperate | Continental | Polar

**Application scale** Element/Material | Component | Façade system

**Type** Insulation | Glazing | Shading | Opening systems | Energy harvesting/generation device | Other

**Purpose** Thermal comfort | Visual performance | Acoustic comfort | Energy management/generation | IAQ | Interaction and users' control | Other

**Adaptive function** Collect and storage | Control | Prevent/Reject | Allow interaction

**Classification** High-performance | Selective | Dynamic façade | Adaptive glazing | Energy generation | Control system

**Operation mode** Passive | Active

**Trigger** Thermal | Electromagnetic | Optical | Air quality | Occupants | Other

**Response input** Extrinsic | Intrinsic | Self-adjusting | Users' control

**Degree of adaptability** On/Off | Gradual

**TRL** 9

**Kuggen (2011)**
Wingårdh Arkitektkontor, Gothenburg (S)

**Fig. 4.22** Kuggen office building. Source: https://commons.wikimedia.org/wiki/File:Kuggen.JPG by Mats Kristoffersen (own work). CC BY 3.0 (https://creativecommons.org/licenses/by/3.0) via Wikimedia Commons

**Introduction** The Kuggen was born to provide optimal and repeated units for offices in the center of Gothenburg. The structure makes use of green building technology on different levels (ventilation, interactive systems for cooling and heating, and effective use of natural light); even the circular floor is designed to minimize the ratio between the amount of the exposed envelope and the building area: Its particular feature of extending its surface by adding two room units per floor allows to shade the windows below. The result is a building with a calculated energy consumption under 55 kWh/m$^2$ annually including all mechanical systems.

**The Smart Feature** The building's outer skin is formed by triangular glazed terracotta prefabricated elements (selected for their longevity), in different shades, in alternation with glazed surface. On the upper floor, a screen with photovoltaic cells movable around the outer perimeter adds shade and electricity, following the sun's path. By equipping the building with motion-activated lighting and ventilation systems, energy is used only where it is really needed.

**Intended use** Residential I Service industry I Specialized

**Building state of development** Concept/Prototype I Real-scale edifice I Under construction

**Climate** Tropical I Dry/Desertic I Temperate I Continental I Polar

**Application scale** Element/Material I Component I Façade system

**Type** Insulation I Glazing I Shading I Opening systems I Energy harvesting/generation device I Other

**Purpose** Thermal comfort I Visual performance I Acoustic comfort I Energy management/generation I IAQ I Interaction and users' control I Other

**Adaptive function** Collect and storage I Control I Prevent/Reject I Allow interaction

**Classification** High-performance I Selective I Dynamic façade I Adaptive glazing I Energy generation I Control system

**Operation mode** Passive I Active

**Trigger** Thermal I Electromagnetic I Optical I Air quality I Occupants I Other

**Response input** Extrinsic I Intrinsic I Self-adjusting I Users' control

**Degree of adaptability** On/Off I Gradual

**TRL** 9

**RMIT's Design School (2012)**
Sean Godsell, Melbourne (AUS)

**Fig. 4.23** (**a**) RMIT whole façade. Source: https://commons.wikimedia.org/wiki/File:Hub_Edge_(14521026881).jpg by Rob Deutscher (own work). CC BY 2.0 (https://creativecommons.org/licenses/by/2.0/deed.en) via Wikimedia commons. (**b**) Building envelope detail. Source: https://commons.wikimedia.org/wiki/File:RMIT_Design_Hub_detail.jpg by Rob Deutscher (own work). CC BY 2.0 (https://creativecommons.org/licenses/by/2.0/deed.en) via Wikimedia commons

**Introduction** The Design Hub of RMIT is intended to accommodate the organic nature of research, constantly evolving, adapting, changing, and growing. The building incorporates strategies of water, waste, and recycling management to provide passive cooling and lower energy consumption coupled with a more desirable thermal comfort alternative to a fully conditioned work environment.

**The Smart Feature** The external envelope, made of a double-glazed inner skin, presents automated operable sun-shading devices that include photovoltaic cells, designed to be easily replaced and especially to contribute to research into the field to be conducted jointly by industry and RMIT. They are able to rotate on a vertical axis to optimize solar radiation, limiting undesired overheating while allowing a good natural light diffusion within confined spaces. The glass cells track the sun via a building computer automation system to help shade and power the building. In sections of the façade, BIPV will be incorporated, manufactured using the same high-performing interlayer, and acting as applied learning and teaching showcase. The whole building façade has the capacity to be upgraded as solar technology evolves and may one day be able to generate enough electricity to run the entire building.

**The technologies** The second skin shading devices are made of 600 mm diameter sandblasted glass disks, which are fixed to an axles grid and in turn fixed to the outer face of a steel cylinder. Twenty-one glass disks and steel cylinders are fixed together in panels of about 1.8 × 4.2 m; each panel is composed of 12 operable disks and 9 fixed disks. At the ground level, all glass disks are fixed.

**Intended use** Residential | Service industry | Specialized

**Building state of development** Concept/Prototype | Real-scale edifice | Under construction

**Climate** Tropical | Dry/Desertic | Temperate | Continental | Polar

**Application scale** Element/Material | Component | Façade system

**Type** Insulation | Glazing | Shading | Opening systems | Energy harvesting/generation device | Other

**Purpose** Thermal comfort | Visual performance | Acoustic comfort | Energy management/generation | IAQ | Interaction and users' control | Other

**Adaptive function** Collect and storage | Control | Prevent/Reject | Allow interaction

**Classification** High-performance | Selective | Dynamic façade | Adaptive glazing | Energy generation | Control system

**Operation mode** Passive | Active

**Trigger** Thermal | Electromagnetic | Optical | Air quality | Occupants | Other

**Response input** Extrinsic | Intrinsic | Self-adjusting | Users' control

**Degree of adaptability** On/Off | Gradual

**TRL** 8

**Smart Material Houses Project: Smart is green (2012)**
zillerplus Architekten und Stadtplaner, Hamburg (DE)

**Fig. 4.24** Smart is Green House in Hamburg. Source: https://commons.wikimedia.org/wiki/File:IBA_Hamburg_Smart_ist_grün.nnw.jpg by NordNordWest (own work). CC BY-SA 3.0 DE (https://creativecommons.org/licenses/by-sa/3.0/de/legalcode) via Wikimedia commons

**Introduction** Project in which the whole envelope is an active part of the building concept; using innovative technology, indeed, it is able to generate power that is fed directly into the house.

**The Smart Feature** The building is able to generate more energy than its use, thanks to the use of smart techniques. In addition, even apartment configuration is smart due to its ability to be modified in a versatile way allowing them to be separated or extended to meet the needs and changing living arrangements of their inhabitants.

**The technologies** The external envelope is made up of three different layers: a vertical garden façade that acts as a heat shield during summer, an insulating glazing layer, and a PCM curtain wall, used as a central storage unit for heat, divided into two tanks. In addition, the outer skin is equipped with photovoltaic panels while solar thermal arrays on the roof convert sunlight into thermal energy.

**Intended use** Residential | Service industry | Specialized

**Building state of development** Concept/Prototype | Real-scale edifice | Under construction

**Climate** Tropical | Dry/Desertic | Temperate | Continental | Polar

**Application scale** Element/Material | Component | Façade system

**Type** Insulation | Glazing | Shading | Opening systems | Energy harvesting/generation device | Other

**Purpose** Thermal comfort | Visual performance | Acoustic comfort | Energy management/generation | IAQ | Interaction and users' control | Other

**Adaptive function** Collect and storage | Control | Prevent/Reject | Allow interaction

**Classification** High-performance | Selective | Dynamic façade | Adaptive glazing | Energy generation | Control system

**Operation mode** Passive | Active

**Trigger** Thermal | Electromagnetic | Optical | Air quality | Occupants | Other

**Response input** Extrinsic | Intrinsic | Self-adjusting | Users' control

**Degree of adaptability** On/Off | Gradual

**TRL** 7

**Smart Material Houses Project: Soft House (2012)**
Kennedy & Violich Architecture, Hamburg (DE)

**Fig. 4.25** Soft House in Hamburg. Source https://www.flickr.com/photos/gunnarries/9191270745/in/photostream/ by Gunnar Ries zwo (own work). CC BY-SA 2.0 (https://creativecommons.org/licenses/by-sa/2.0/) via Flickr

**Introduction** The Soft House is part of the competition for the International Bau Austellung (IBA) of 2012. It represents, like its competitors, a novel model for low-carbon construction that can be customized to meet needs of end users. It is based on a flexible living concept that integrates architecture, smart technologies, and clean energy infrastructures.

**The Smart Feature** The façade, covered with a dynamic textile membrane, is moveable according to sunlight variations; photovoltaic cells are incorporated within it to exploit the sun for producing energy. In addition, movable transparent curtains—equipped with LED lighting systems—are adopted inside the apartments to change their layout on demand, replacing traditionally steady interior elements; their operation is guaranteed by the electricity generated by the membrane façade.

**The technologies** The building's structure consists of wood panels and flexible solar nanomaterials embedded in a light, smart textile that bends to reach for the best angle to harvest solar energy.

**Intended use** Residential | Service industry | Specialized

**Building state of development** Concept/Prototype | Real-scale edifice | Under construction

**Climate** Tropical | Dry/Desertic | Temperate | Continental | Polar

**Application scale** Element/Material | Component | Façade system

**Type** Insulation | Glazing | Shading | Opening systems | Energy harvesting/generation device | Other

**Purpose** Thermal comfort | Visual performance | Acoustic comfort | Energy management/generation | IAQ | Interaction and users' control | Other

**Adaptive function** Collect and storage | Control | Prevent/Reject | Allow interaction

**Classification** High-performance | Selective | Dynamic façade | Adaptive glazing | Energy generation | Control system

**Operation mode** Passive | Active

**Trigger** Thermal | Electromagnetic | Optical | Air quality | Occupants | Other

**Response input** Extrinsic | Intrinsic | Self-adjusting | Users' control

**Degree of adaptability** On/Off | Gradual

**TRL** 7

**South Australian Health and Medical Research Institute—SAHMRI (2012)**
Woods Bagot, Adelaide (AUS)

**Fig. 4.26** SAHMRI. Source https://www.flickr.com/photos/danthemanocallaghan/15522915098/in/photostream/ by Dan O'Cker (own work). CC BY-SA 2.0 (https://creativecommons.org/licenses/by-sa/2.0/) via Flickr

**Introduction** The South Australian Health and Medical Research Institute (SAHMRI) is designed to become a world-class center of research excellence in medicine, creating synergies among researchers and clinicians and integrating research into practice. Therefore, even the architecture has been conceived to allow visual connections between inside and outside and floors as well.

**The Smart Feature** The particular façade of the building, inspired by the skin of a pine cone, is able to adapt and respond to the external environment, thanks to the triangulated movable structure that acts as an articulated sun-shading device. Its specific design has been developed following intensive environmental analysis and parametric modeling tools to integrate environmental and formal requirements into such a system.

**The technologies** The sun-shading system is made of a free-hanging metallic grid placed on each glass panel of the curtain wall, with every shade designed site-specifically to modulate the amount of light needed inside the building as determined by computer modeling.

**Intended use** Residential | Service industry | Specialized

**Building state of development** Concept/Prototype | Real-scale edifice | Under construction

**Climate** Tropical | Dry/Desertic | Temperate | Continental | Polar

**Application scale** Element/Material | Component | Façade system

**Type** Insulation | Glazing | Shading | Opening systems | Energy harvesting/generation device | Other

**Purpose** Thermal comfort | Visual performance | Acoustic comfort | Energy management/generation | IAQ | Interaction and users' control | Other

**Adaptive function** Collect and storage | Control | Prevent/Reject | Allow interaction

**Classification** High-performance | Selective | Dynamic façade | Adaptive glazing | Energy generation | Control system

**Operation mode** Passive | Active

**Trigger** Thermal | Electromagnetic | Optical | Air quality | Occupants | Other

**Response input** Extrinsic | Intrinsic | Self-adjusting | Users' control

**Degree of adaptability** On/Off | Gradual

**TRL** 9

**Windswept at Randall Museum (2012)**
Charles Sowers, San Francisco, CA (USA)

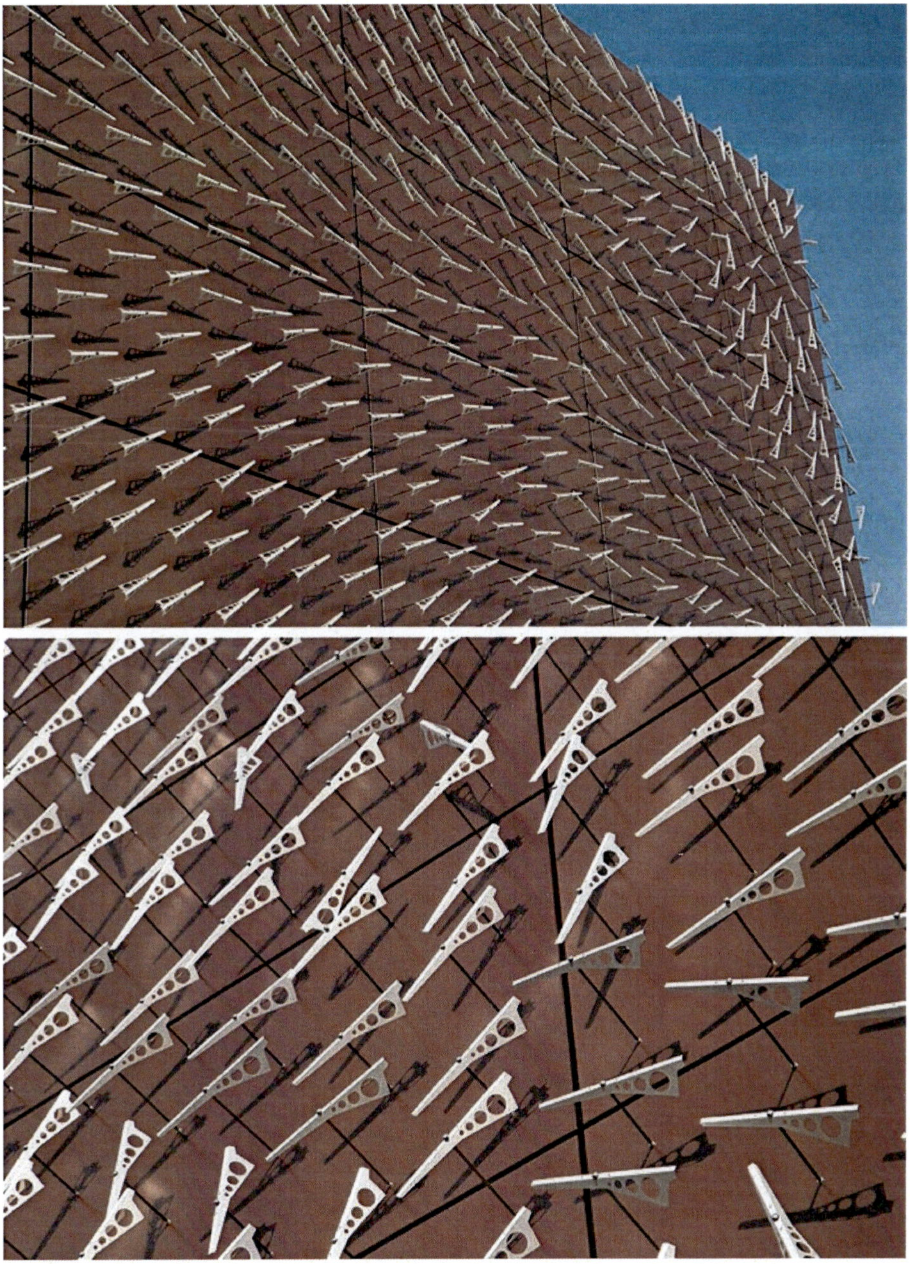

**Fig. 4.27** Details of the Windswept installation on the Randall Museum façades in SF. Source: https://www.charlessowers.com/new-page-1 by permission of Charles Sowers

**Introduction** Windswept is an installation on the Randall Museum façade developed by Charles Sowers with the aim to show the interactions between the wind and the environment.

**The Smart Feature** This wind-driven kinetic façade would inspire to think about the wind as a potential energy source, sensitizing people about energy consumption and conferring to the building an ever-changing aspect due to the multiple interactions it has with the wind.

**The technologies** Made of 612 anodized aluminum arrows, mounted parallel to the façade following a grid, the installation shows the air movements through the façade of the museum. It consists of over 500 freely rotating directional arrows that use the power of the wind to move.

**Intended use** Residential | Service industry | <u>Specialized</u>

**Building state of development** Concept/Prototype | <u>Real-scale edifice</u> | Under construction

**Climate** Tropical | Dry/Desertic | <u>Temperate</u> | Continental | Polar

**Application scale** <u>Element/Material</u> | Component | Façade system

**Type** Insulation | Glazing | Shading | Opening systems | Energy harvesting/generation device | <u>Other</u>

**Purpose** Thermal comfort | Visual performance | Acoustic comfort | Energy management/generation | IAQ | <u>Interaction and users' control</u> | Other

**Adaptive function** Collect and storage | Control | Prevent/Reject | <u>Allow interaction</u>

**Classification** High-performance | Selective | <u>Dynamic façade</u> | Adaptive glazing | Energy generation | Control system

**Operation mode** <u>Passive</u> | Active

**Trigger** Thermal | Electromagnetic | Optical | Air quality | Occupants | <u>Other</u>

**Response input** Extrinsic | <u>Intrinsic</u> | Self-adjusting | Users' control

**Degree of adaptability** On/Off | <u>Gradual</u>

**TRL** 9

**One Ocean Thematic Pavilion EXPO (2012)**
SOMA Lima, Yeosu-Si (ROK)

**Fig. 4.28** One Ocean Expo 2012 theme pavilion. Source: https://commons.wikimedia.org/wiki/File:Expo2012_Theme_pavilion.JPG by hyolee2 (own work). CC BY-SA 3.0 (https://creativecommons.org/licenses/by/3.0/deed.en) via Wikimedia Commons

**Introduction** One Ocean thematic pavilion for the EXPO 2012 was inspired by the Expo's theme, which was "The Living Ocean and Coast." Thanks to its design and organic surfaces indeed, it allows the visitor to fully experience the ocean. Thanks to its particular aesthetic indeed, it fully embodies the Expo's theme in architecture, transforming it into a multilayered experience according to a biomimetic approach.

**The Smart Feature** The building's roof turns into a kinetic media façade that allows the building to "breath." The façade, made of glass fiber reinforced polymers (GFRP) lamellas able to move, was inspired by a project that investigates how biological mechanisms can be applied on an architectural scale. It is composed of 108 vertical lamellas (that span 140 m, with elevation from 3 to 13 m in height), individually controlled, which allows regulating the input of indoor solar energy while enhancing natural ventilation by capturing and directing winds through the building during intermediate seasons.

**The tchnologies** The material that shapes the lamellas, a GFRP, is high in tensile strength with low bending stiffness, allowing for repeated and reversible elastic deformations. The visual effect of the opening movements is enhanced by LEDs fitted into the inner side of the lamellas.
Photovoltaic panels are integrated into the roof landscape to generate solar electricity, providing about two-thirds of the energy consumed by the building systems over the year.

**Intended use** Residential | Service industry | <u>Specialized</u>

**Building state of development** Concept/Prototype | <u>Real-scale edifice</u> | Under construction

**Climate** Tropical | Dry/Desertic | Temperate | <u>Continental</u> | Polar

**Application scale** Element/Material | Component | <u>Façade system</u>

**Type** Insulation | Glazing | Shading | <u>Opening systems</u> | Energy harvesting/generation device | Other

**Purpose** Thermal comfort | <u>Visual performance</u> | Acoustic comfort | Energy management/generation | IAQ | Interaction and users' control | Other

**Adaptive function** Collect and storage | <u>Control</u> | Prevent/Reject | Allow interaction

**Classification** High-performance | Selective | <u>Dynamic façade</u> | Adaptive glazing | Energy generation | Control system

**Operation mode** Passive | <u>Active</u>

**Trigger** Thermal | <u>Electromagnetic</u> | Optical | Air quality | Occupants | Other

**Response input** Extrinsic | <u>Intrinsic</u> | Self-adjusting | Users' control

**Degree of adaptability** On/Off | <u>Gradual</u>

**TRL** 8

**Al Bahar Towers (2013)**
AEDAS UK, Abu Dhabi (UAE)

**Fig. 4.29** Al Bahar Towers: Detail of the dynamic façades. Source https://www.flickr.com/photos/inhabitat/12330902845/in/photostream/ by Inhabitat. CC BY-NC-ND 2.0 (https://creativecommons.org/licenses/by-nc-nd/2.0/) via Flickr

**Introduction** Honored with the Tall Building Innovation Award, the towers, headquarters of the Abu Dhabi Investment Council, make their special building envelope system their distinguishing feature. Totally glazed, their ability to adapt in an extreme climatic context such as the Emirates' desert is guaranteed by the peculiar shading systems, fully automated and able to be activated when the external conditions change, controlling, in a selective way, the solar radiation without compromising the outside view.

**The Smart Feature** The responsive façade, inspired from the traditional Islamic shading device, reacts in direct response to the sun's movements. Thanks to these smart devices, indeed, it was possible to create a fully glazed building without recurring to heavily tinted glazing that could have affected natural lighting within confined spaces.

**The technologies** This system, made by over 1000 hexagonal PTFE elements inspired by the typical shape of the Arab *mashrabiya* (each tower counts 1049 shading devices, each weighing about 1.5 tones), reacts according to the external temperature, contracting itself when it goes down. Its dimension and, in particular, the definition of the geometry of the single inserts were realized; thanks to a parametric descriptor that simulated its behavior during a daily solar cycle. The screen operates as a curtain wall separated from the buildings bearing structure; each triangle is programmed to respond to the movement of the sun; in the evening, all the screens will close, completely changing the appearance of the building. The shape of the building in plan and elevation led to 22 different variations in its geometry.

**The Performance** In this building, the presence of well-established construction elements, such as shading systems reinterpreted according to local traditions, has allowed achieving a reduction in energy consumption of about 50% and solar gain of about 80%.

**Intended use** Residential | Service industry | Specialized

**Building state of development** Concept/Prototype | Real-scale edifice | Under construction

**Climate** Tropical | Dry/Desertic | Temperate | Continental | Polar

**Application scale** Element/Material | Component | Façade system

**Type** Insulation | Glazing | Shading | Opening systems | Energy harvesting/generation device | Other

**Purpose** Thermal comfort | Visual performance | Acoustic comfort | Energy management/generation | IAQ | Interaction and users' control | Other

**Adaptive function** Collect and storage | Control | Prevent/Reject | Allow interaction

**Classification** High-performance | Selective | Dynamic façade | Adaptive glazing | Energy generation | Control system

**Operation mode** Passive | Active

**Trigger** Thermal | Electromagnetic | Optical | Air quality | Occupants | Other

**Response input** Extrinsic | Intrinsic | Self-adjusting | Users' control

**Degree of adaptability** On/Off | Gradual

**TRL** 8

**Hanwha Headquarters Remodeling (2013)**
UNStudio, Seoul (ROK)

**Fig. 4.30** Details of the Hanwha Headquarters PV Façade. Photo Rohspace©. Courtesy of UNStudio

**Introduction** The façade is part of a refurbishment project of the building aimed at renovating the aspect of the Hanwha Headquarters; located in Cheonggyecheon in Seoul, it aims at reflecting the role of the company as leading environmental technology providers in the world.
Based on environmental considerations, a conscious decision was made to remodel the existing building, rather than for demolition and new build. Furthermore, employing an efficient construction method, the firm was also able to ensure that the employees could continue to work in the building throughout the renovation.

**The Smart Feature** A completely new, energy-generating façade with integrated solar panels. The novel façade is made of clear insulated glass with aluminum framing, whose position and geometry were sized and calculated in relation to the movement of the sun to ensure users' comfort and reducing energy consumption; by varying the placement of the façade panels, a variety of program-related openings are created. PV cells are placed on the opaque panels on the south/southeast façade at the open zones where there is an optimal amount of direct sunlight. Furthermore, PV panels are angled in the areas of the façade where energy from the sun can best be harvested.

**Intended use** Residential | Service industry | Specialized

**Building state of development** Concept/Prototype | Real-scale edifice | Under construction

**Climate** Tropical | Dry/Desertic | Temperate | Continental | Polar

**Application scale** Element/Material | Component | Façade system

**Type** Insulation | Glazing | Shading | Opening systems | Energy harvesting/generation device | Other

**Purpose** Thermal comfort | Visual performance | Acoustic comfort | Energy management/generation | IAQ | Interaction and users' control | Other

**Adaptive function** Collect and storage | Control | Prevent/Reject | Allow interaction

**Classification** High-performance | Selective | Dynamic façade | Adaptive glazing | Energy generation | Control system

**Operation mode** Passive | Active

**Trigger** Thermal | Electromagnetic | Optical | Air quality | Occupants | Other

**Response input** Extrinsic | Intrinsic | Self-adjusting | Users' control

**Degree of adaptability** On/Off | Gradual

**TRL** 7

**Smart Material House BIQ (2013)**
Arup and Splitterwerk Architects, Hamburg (DE)

**Fig. 4.31** Smart Material House BIQ in Hamburg. Source: https://commons.wikimedia.org/wiki/File:IBA_Hamburg_BIQ_(2).nnw.jpg by NordNordWest (own work). CC BY-SA 3.0 (https://creativecommons.org/licenses/by-sa/3.0/de/legalcode) via Wikimedia commons

**Introduction** The building with Bio-Intelligent Quotient (BIQ) is a pilot project with a bioreactive façade, presented at the International Building Exhibition in Hamburg in 2013 as the first totally passive building in the world powered by algae. It is a solid cubic structure of stonework and concrete that combines adaptable structural design (different layout typologies of apartments are available in response to contemporary demands) with smart technologies and building materials.

**The Smart Feature** The use of material largely available in nature, such as renewable energy source; integrating additional features to the production system, such as dynamic shading—determined by the variable growth of algae as a function of incident solar radiation—and thermal and acoustic insulation; giving life to an organism strongly characterized by the technology employed, also from an aesthetic point of view; and potentially endowed with significant consequences also on the surrounding environment as a means to reduce $CO_2$ emissions through buildings' façades.

**The technologies** The system, installed on the southeast and south-west building façades (200 m$^2$ approximately), is made up of 129 sun-tracking reactor modules consisting of a double-glazed window (with argon in the air-gap), mounted on a metal frame and separated by a cavity of about 24 lt, filled with water and suitable for the culture medium needed for the proliferation of micro-algae. The algae's growth, stimulated by the incident solar radiation and by the addition of $CO_2$ as a nutrient, allows the production of heat (used for heating the rooms and for the production of domestic hot water or stored for further uses) and biomass (about 30 KWh/m$^2$ year), collected by means of a separator and placed in a container at a controlled temperature to be further used for biogas production. The whole system operation is guaranteed by a centralized management device that manages all the processes necessary to operate the bioreactor façade and to fully integrate it with the energy management system of the building.

**The Performance:** Due to the sunlight and a constant turbulence that avoids algae aggregation, the production of heat is about 38% of its efficiency, compared to 60–65% of a conventional solar thermal, while biomass production is about 10% of efficiency in comparison to 12–15% of a conventional PV.

**Intended use** <u>Residential</u> | Service industry | Specialized

**Building state of development** <u>Concept/Prototype</u> | Real-scale edifice | Under construction

**Climate** Tropical | Dry/Desertic | <u>Temperate</u> | Continental | Polar

**Application scale** Element/Material | <u>Component</u> | Façade system

**Type** Insulation | Glazing | Shading | Opening systems | <u>Energy harvesting/generation device</u> | Other

**Purpose** Thermal comfort | Visual performance | Acoustic comfort | Energy management/generation | IAQ | Interaction and users' control | <u>Other</u>

**Adaptive function** <u>Collect and storage</u> | Control | Prevent/Reject | Allow interaction

**Classification** <u>High-performance</u> | Selective | Dynamic façade | Adaptive glazing | Energy generation | Control system

**Operation mode** <u>Passive</u> | Active

**Trigger** Thermal | <u>Electromagnetic</u> | Optical | Air quality | Occupants | Other

**Response input** Extrinsic | <u>Intrinsic</u> | Self-adjusting | Users' control

**Degree of adaptability** On/Off | <u>Gradual</u>

**TRL** 7

**Torre de Especialidades, Hospital Manuel Gea Gonzales (2013)**
elegant embellishments, Mexico city (MEX)

**Fig. 4.32** (**a**) View of the Hospital façade from the street. Source: https://web.archive.org/web/20161030011416/http://www.panoramio.com/photo/101612732 by ArCaRa (own work). CC BY 3.0 (https://creativecommons.org/licenses/by/3.0/deed.en) via Panoramio. (**b**) Detail of the façade's ceramic modules. Source: https://commons.wikimedia.org/wiki/File:Doble_fachada_de_la_Torre_de_especialidades_del_Hospital_General_Dr._Gea_González.jpg by Linda Oviedo (own work). CC BY-SA 4.0 (https://creativecommons.org/licenses/by-sa/4.0/deed.en) via Wikimedia commons

**Introduction** The Torre de Especialidades is an addition to the Hospital Manuel Gea Gonzales, located in the southern Tlalpan neighborhood of Mexico City. Built as part of a government project to improve the city's health infrastructure (called ProAire initiative), it is designed to transform air pollutants into harmless chemicals, improving surrounding air quality (an aspect of particular importance in an area full of hospital patients).

**The Smart Feature** The building façade (approximately of about 2500 $m^2$) is made up of a new type of tile that can neutralize the surrounding air pollutants; thanks to the special paint, within which the façade's modules are treated, made from a titanium dioxide-based pigment. When daylight reaches the tiles' surface, the titanium dioxide triggers a chemical reaction between the treated surface and the smog in the air, reducing air pollution by breaking down smog into safer chemicals.

**The technologies** Inspired by fractals in nature, the undulating shapes of the façade's modules are designed to better distribute pollutants across the active surface, slowing wind speed and creating turbulence. Moreover, the particular geometry of the tiles, obtained through lightweight thermoformed plastic panels, allows shadowing the inside of the building, helping to maintain proper internal comfort conditions and at the same time lowering the electric energy needed for conditioning.

**The Performance** Based on third-party testing of the material employed for the realization of this façade, it has been estimated that it allows to neutralize roughly the same amount of smog produced each day by about 1000 vehicles in Mexico City. Furthermore, performing its depolluting action, the pigment that covers the tiles remains unchanged, which means that such a system can keep purifying the air for as long as a decade.

**Intended use** Residential | Service industry | <u>Specialized</u>

**Building state of development** Concept/Prototype | <u>Real-scale edifice</u> | Under construction

**Climate** Tropical | Dry/Desertic | <u>Temperate</u> | Continental | Polar

**Application scale** Element/Material | <u>Component</u> | Façade system

**Type** Insulation | Glazing | Shading | Opening systems | Energy harvesting/generation device | <u>Other</u>

**Purpose** Thermal comfort | Visual performance | Acoustic comfort | Energy management/generation | <u>IAQ</u> | Interaction and users' control | Other

**Adaptive function** Collect and storage | Control | <u>Prevent/Reject</u> | Allow interaction

**Classification** <u>High-performance</u> | Selective | Dynamic façade | Adaptive glazing | Energy generation | Control system

**Operation mode** <u>Passive</u> | Active

**Trigger** Thermal | Electromagnetic | <u>Optical</u> | Air quality | Occupants | Other

**Response input** Extrinsic | <u>Intrinsic</u> | Self-adjusting | Users' control

**Degree of adaptability** <u>On/Off</u> | Gradual

**TRL** 8

### Festo's Automation Centre (2014)

Ulrich & Jens Jaschek Architects, Esslingen (DE)

**Fig. 4.33** Festo Hochaus, Esslingen. Source: https://commons.wikimedia.org/wiki/File:Festo_Hochhaus_Esslingen1.jpg by Silesia711 (own work). CC BY-SA 4.0 (https://creativecommons.org/licenses/by-sa/4.0/deed.en) via Wikimedia commons

**Introduction** Located in the south of Germany, Festo's Automation Centre presents a smart automated glass façade (of about 8500 m$^2$) that fulfills Germany's strict energy standards.

**The Smart Feature** The completely glazed façade: dynamic, flow-optimized, pneumatic, and with solar energy recovery. Indeed, it is designed as an internally ventilated façade (with a ventilation channel between the screen and the exterior envelope) but with a dynamically adaptable solar energy transmittance, thanks to electrochromic glass panels.

Building Automation Systems that cover all areas of the technical infrastructures (e.g. heating, ventilation, air conditioning, lighting, etc.) provide for automatic control. In addition, electrochromic glazing ensures reduced light and heat, preserving the unobstructed view and natural ventilation, which is possible at any time, as the windows can be opened manually.

A comprehensive measuring and monitoring system is integrated to carry out a further optimization of the operating strategies for the system engineering after evaluation.

**The Performance** The system thus obtained is extremely efficient: Even at an outside temperature of over 35 °C, the indoor temperature remains constant and never exceeds 26 °C.

**Intended use** Residential | Service industry | Specialized

**Building state of development** Concept/Prototype | Real-scale edifice | Under construction

**Climate** Tropical | Dry/Desertic | Temperate | Continental | Polar

**Application scale** Element/Material | Component | Façade system

**Type** Insulation | Glazing | Shading | Opening systems | Energy harvesting/generation device | Other

**Purpose** Thermal comfort | Visual performance | Acoustic comfort | Energy management/generation | IAQ | Interaction and users' control | Other

**Adaptive function** Collect and storage | Control | Prevent/Reject | Allow interaction

**Classification** High-performance | Selective | Dynamic façade | Adaptive glazing | Energy generation | Control system

**Operation mode** Passive | Active

**Trigger** Thermal | Electromagnetic | Optical | Air quality | Occupants | Other

**Response input** Extrinsic | Intrinsic | Self-adjusting | Users' control

**Degree of adaptability** On/Off | Gradual

**TRL** 9

**May-September, Eskenazi Hospital façade (2014)**
Rob Ley Studio, Indianapolis (USA)

**Fig. 4.34** Eskenazi Hospital façade. Source: https://commons.wikimedia.org/wiki/File:YH48AINJxjw.jpg by Rockinrollblues (own work). CC BY-SA 4.0 (https://creativecommons.org/licenses/by-sa/4.0/deed.en) via Wikimedia commons

**Introduction** Commonly identified with the façade of the Indianapolis hospital, this dynamic envelope is actually the cladding of the multistory car park that faces it; originated to transform a generally unappreciated infrastructure—as a parking—into an interactive terrain.

**The Smart Feature** The chromatic variations of the façade, coupled with the different angles of the panels that made it up, offer a dynamic visual experience, implementing performance features, such as shielding and ventilation functions.

**The technologies** Made using about 7000 panels in aluminum and steel, the façade has been designed according to the direction of the prevailing winds on the site as well as to users' points of view; thanks to a computer-aided system that simulated the mutual arrangement of the individual panels, thus optimizing the final image obtained. Therefore, its main feature is the three-dimensionality, obtained through the mobility of the elements that compose it in users' perception and emphasized by the different colors on their two sides rather than by means of actual kinetic components.

**Intended use** Residential | Service industry | Specialized

**Building state of development** Concept/Prototype | Real-scale edifice | Under construction

**Climate** Tropical | Dry/Desertic | Temperate | Continental | Polar

**Application scale** Element/Material | Component | Façade system

**Type** Insulation | Glazing | Shading | Opening systems | Energy harvesting/generation device | Other

**Purpose** Thermal comfort | Visual performance | Acoustic comfort | Energy management/generation | IAQ | Interaction and users' control | Other

**Adaptive function** Collect and storage | Control | Prevent/Reject | Allow interaction

**Classification** High-performance | Selective | Dynamic façade | Adaptive glazing | Energy generation | Control system

**Operation mode** Passive | Active

**Trigger** Thermal | Electromagnetic | Optical | Air quality | Occupants | Other

**Response input** Extrinsic | Intrinsic | Self-adjusting | Users' control

**Degree of adaptability** On/Off | Gradual

**TRL** 9

## SDU Campus (2014)

Henning Larsen Architects, Kolding (DK)

**Fig. 4.35** (**a**) SDU Campus: view from the street. Source: https://commons.wikimedia.org/wiki/File:Syddansk_universitet.Campus_Kolding.Denmark.2014_(38).JPG (**b**) Detail of the shading movable system. Source: https://commons.wikimedia.org/wiki/File:Syddansk_universitet.Campus_Kolding.Denmark.2014_(34).JPG by S. Juhl (own work). CC0 1.0 (https://creativecommons.org/publicdomain/zero/1.0/deed.en) via Wikimedia commons

**Introduction** The new campus of the Syddansk Universitet Design School has a triangular shape and created a new significant landmark in Kolding.

**The Smart Feature** The climate-responsive kinetic façade, obtained through dynamic solar shading, adjusts to outdoor climate conditions and users' patterns as well, providing optimal daylighting during the whole day.
**The technologies** The shading system consists of approximately 1600 triangular elements of perforated steel, mounted on the façade in a way that allows them to move according to the changing of daylight levels. Their rotation defines shapes and transparencies always different; even the perforation is articulated following a circular pattern, and it was studied to optimize the ration among users' view toward the outside and façade's shading function. This system is equipped with a sensor that continuously measures light and heat levels and that regulates the shutters accordingly by means of a motor.

**Intended use** Residential | Service industry | <u>Specialized</u>

**Building state of development** Concept/Prototype | <u>Real-scale edifice</u> | Under construction

**Climate** Tropical | Dry/Desertic | <u>Temperate</u> | Continental | Polar

**Application scale** Element/Material | <u>Component</u> | Façade system

**Type** Insulation | Glazing | <u>Shading</u> | Opening systems | Energy harvesting/generation device | Other

**Purpose** Thermal comfort | <u>Visual performance</u> | Acoustic comfort | Energy management/generation | IAQ | Interaction and users' control | Other

**Adaptive function** Collect and storage | <u>Control</u> | Prevent/Reject | Allow interaction

**Classification** High-performance | <u>Selective</u> | Dynamic façade | Adaptive glazing | Energy generation | Control system

**Operation mode** Passive | <u>Active</u>

**Trigger** Thermal | <u>Electromagnetic</u> | Optical | Air quality | Occupants | Other

**Response input** <u>Extrinsic</u> | Intrinsic | Self-adjusting | Users' control

**Degree of adaptability** On/Off | <u>Gradual</u>

**TRL** 9

**The W.I.N.D. House (2014)**
UNStudio, North-Holland (NL)

**Fig. 4.36** The W.I.N.D. House. Photos: Fedde de Weert©. Courtesy of UNStudio

**Introduction** Described as the "home of the future" due to the interaction that occurs through its systems and appliances—thanks to ICTs—thus resulting in energy consumption reduction, safety, and security, the W.I.N.D. House incorporates both integrated sustainable solutions and home automations regulating and maximizing the effects of its surrounding landscape. In addition, flexible layouts are available to respond to family's changing needs.
**The Smart Feature** The organization of the house, defined by its external conditions, and the complete control of all the home's appliances, including electrical systems, solar panels, and mechanical installations. Automation and connectivity make it possible for homes to integrate information technologies through which systems and appliances are able to communicate in an integrated way. Also, the home responds to changes in today's lifestyles, thanks to incorporated flexible floor plans that allow for diversity in function through the years.

**The technologies** The complete control of this Smart Home is possible by a central touch-screen interface located in the living area; furthermore, remote control is possible by independent devices via LAN connection. The comprehensive home automation system enables integrated control of the electrical systems, including solar panels and mechanical installations.

**Intended use** Residential | Service industry | Specialized

**Building state of development** Concept/Prototype | Real-scale edifice | Under construction

**Climate** Tropical | Dry/Desertic | Temperate | Continental | Polar

**Application scale** Element/Material | Component | Façade system

**Type** Insulation | Glazing | Shading | Opening systems | Energy harvesting/generation device | Other

**Purpose** Thermal comfort | Visual performance | Acoustic comfort | Energy management/generation | IAQ | Interaction and users' control | Other

**Adaptive function** Collect and storage | Control | Prevent/Reject | Allow interaction

**Classification** High-performance | Selective | Dynamic façade | Adaptive glazing | Energy generation | Control system

**Operation mode** Passive | Active

**Trigger** Thermal | Electromagnetic | Optical | Air quality | Occupants | Other

**Response input** Extrinsic | Intrinsic | Self-adjusting | Users' control

**Degree of adaptability** On/Off | Gradual

**TRL** 8

**Umbrella façade at Madrid pavilion (2014)**
3GATTI Architecture Studio, UBPA, Shangai Expo Area (China)

**Fig. 4.37** Rendering of the Madrid pavilion with façade's detail. Courtesy of 3GATTI Architecture Studio

**Introduction** Novel concept of façade developed for the Madrid Pavilion after its conversion into a retail and office complex once the Expo was concluded. The new façade aimed at replacing the bamboo louvers that originally surround the glazed walls of the building.

**The Smart Feature** The shading system, inspired by the Chinese culture, is adjustable on-demand by occupants to control light levels within the building. When the umbrellas are fully opened, the façade is completely flat so that most of the sunlight and strong wind are blocked. If the umbrellas are opened, the light is able to come fully inside, and the umbrella sticks, with an aerodynamic shape not resistant to the wind, preventing structural tensions in case of typhoons.

**The technologies** Starting from the Chinese tradition to repair oneself with umbrellas on sunny days, the architects transferred this concept to the sun-shading devices designed for covering the glazed building's façades. Each "module" is individually controllable by a pulley: the mechanical parts are made of stainless steel, the frames of aluminum, and the external surface a thin perforated Corten.

**Intended use** Residential | Service industry | <u>Specialized</u>

**Building state of development** <u>Concept/Prototype</u> | Real-scale edifice | Under construction

**Climate** Tropical | Dry/Desertic | Temperate | <u>Continental</u> | Polar

**Application scale** Element/Material | <u>Component</u> | Façade system

**Type** Insulation | Glazing | <u>Shading</u> | Opening systems | Energy harvesting/generation device | Other

**Purpose** Thermal comfort | <u>Visual performance</u> | Acoustic comfort | Energy management/generation | IAQ | Interaction and users' control | Other

**Adaptive function** Collect and storage | <u>Control</u> | Prevent/Reject | Allow interaction

**Classification** High-performance | <u>Selective</u> | Dynamic façade | Adaptive glazing | Energy generation | Control system

**Operation mode** Passive | <u>Active</u>

**Trigger** Thermal | Electromagnetic | Optical | Air quality | <u>Occupants</u> | Other

**Response input** Extrinsic | Intrinsic | Self-adjusting | <u>Users' control</u>

**Degree of adaptability** On/Off | <u>Gradual</u>

**TRL** 8

**The Breathing Skins, Showroom (2015)**
Tobias Becker Studio, Mandelbachtal (DE)

**Fig. 4.38** The Breathing skin. Courtesy of Ing. Tobias Becker

**Introduction** Breathing skins, invented and patented by Tobias Becker during his architecture studies, tried to conceive the envelope as an adaptive media of exchange, able to act as a real filter between inside and outside, reducing the barriers between the two environments.

**The Smart Feature** The breathing skin is based on a novel concept of responsive architecture that conceives the indoor as a part of the natural environment. Its layout is determined by six circular segments with different sizes and can be subdivided into two subareas: one with opaque air channels and one with translucent air channels. To achieve the appearance of a continuous skin, all constructive parts are assembled by extruded glass (polycarbonate) while the bearing structure is wooden constructed.

**The technologies** The Breathing Skins is constituted by a set of channels (about 140 on every square meters for a total amount around 2800 air channels) that can be pneumatically closed after the application of air pressure within the façade. On every square meter, 140 air channels are controlled without any visible technical installations. This deformation process, operated by a pneumatic actuator, requires minimal energetic input. The more the pneumatic muscles open, the more the appearance of the façade changes. The façade elements, separated into 14 curved panels, are built as sandwich construction (the pneumatic muscles are joined reversibly between two glass plates) with a weight lower than 11 kg/m$^2$.

**Intended use** Residential | Service industry | <u>Specialized</u>

**Building state of development** <u>Concept/Prototype</u> | Real-scale edifice | Under construction

**Climate** Tropical | Dry/Desertic | <u>Temperate</u> | Continental | Polar

**Application scale** Element/Material | Component | <u>Façade system</u>

**Type** <u>Insulation</u> | Glazing | Shading | Opening systems | Energy harvesting/generation device | Other

**Purpose** <u>Thermal comfort</u> | Visual performance | Acoustic comfort | Energy management/generation | IAQ | Interaction and users' control | Other

**Adaptive function** Collect and storage | Control | <u>Prevent/Reject</u> | Allow interaction

**Classification** <u>High-performance</u> | Selective | Dynamic façade | Adaptive glazing | Energy generation | Control system

**Operation mode** Passive | <u>Active</u>

**Trigger** Thermal | Electromagnetic | Optical | Air quality | Occupants | <u>Other</u>

**Response input** <u>Extrinsic</u> | Intrinsic | Self-adjusting | Users' control

**Degree of adaptability** <u>On/Off</u> | Gradual

**TRL** 7

**Shangai Natural History Museum (2015)**
Perkins and Will, Shangai (RPC)

**Fig. 4.39** Detail of the museum's courtyard façade. Source: https://commons.wikimedia.org/wiki/File:Shanghai_Natural_History_Museum_(2015)_-_05.JPG by Another Believer (own work). CC BY-SA 4.0 (https://creativecommons.org/licenses/by-sa/4.0/deed.en) via Wikimedia Commons

**Introduction** The museum, conceived as an organic shape to contemplate the human relationship with nature, stands for most below grade and is born to replace the original Natural History Museum, improving its exhibition ability. It was inspired by the nautilus shell, one of the purest geometric forms found in nature, and it is a bioclimatic building whose energy features are fully part of exhibits. Furthermore, every part of the museum is open to human interaction.

**The Smart Feature** It consists of several environmental design solutions that depict natural elements across building's façades, such as the central cell wall, the east living wall, and the northern stone wall. Particularly, the cell wall—the museum's distinctive feature—can be considered an intelligent skin due to its capacity of responding to the sun, maximizing daylight and minimizing solar gain. Moreover, a pond—which collects water from the museum's vegetated roof—regulates indoor temperature, together with geothermal wells.

**The technologies** The wall is organized in an elliptical cone-shaped envelope, composed of three layers; each of them has its own geometrical pattern and organic form. At the core, there is the main layer, then the structural cell layer—which emphasizes the organic cells as structural building blocks of nature—while as the inner layer, there is the actual waterproof envelope of the building, formed by the glass and aluminum mullion curtain wall. Finally, an outer layer provides a solar screen, emulating the cellular building block of all life forms and performs like traditional Chinese window screens.

**Intended use** Residential | Service industry | Specialized

**Building state of development** Concept/Prototype | Real-scale edifice | Under construction

**Climate** Tropical | Dry/Desertic | Temperate | Continental | Polar

**Application scale** Element/Material | Component | Façade system

**Type** Insulation | Glazing | Shading | Opening systems | Energy harvesting/generation device | Other

**Purpose** Thermal comfort | Visual performance | Acoustic comfort | Energy management/generation | IAQ | Interaction and users' control | Other

**Adaptive function** Collect and storage | Control | Prevent/Reject | Allow interaction

**Classification** High-performance | Selective | Dynamic façade | Adaptive glazing | Energy generation | Control system

**Operation mode** Passive | Active

**Trigger** Thermal | Electromagnetic | Optical | Air quality | Occupants | Other

**Response input** Extrinsic | Intrinsic | Self-adjusting | Users' control

**Degree of adaptability** On/Off | Gradual

**TRL** 9

**Forti Holding | Centro Direzionale "Umberto Forti" (2016)**
ATIProject, Pisa, (IT)

**Fig. 4.40** Forti Holding, building view (**a**) and façade detail (**b**). Source: https://atiproject.com/project/umberto-forti-headquarters/. Courtesy of ATIProject

**Introduction** The building is considered as an example of green architecture resulting in extremely high-energetic performance. In its design, opaque and transparent surfaces have been taken carefully in account according to their specific size and orientation.

**The Smart Feature** The smart building envelope system is intended as a mix of technological solutions able to meet the current requirements and needs of the users as well.

**The technologies** The building envelope is made up of a ventilated façade and a totally glazed façade, with a sunscreen system to filter light entering within confined spaces. The south façade is covered with solar panels, and it has been dimensioned to fully meet the building's energetic needs. Thanks to the special technological solutions adopted, the envelope becomes the first system of the building able to significantly limit its consumption.

**Intended use** Residential | Service industry | Specialized

**Building state of development** Concept/Prototype | Real-scale edifice | Under construction

**Climate** Tropical | Dry/Desertic | Temperate | Continental | Polar

**Application scale** Element/Material | Component | Façade system

**Type** Insulation | Glazing | Shading | Opening systems | Energy harvesting/generation device | Other

**Purpose** Thermal comfort | Visual performance | Acoustic comfort | Energy management/generation | IAQ | Interaction and users' control | Other

**Adaptive function** Collect and storage | Control | Prevent/Reject | Allow interaction

**Classification** High-performance | Selective | Dynamic façade | Adaptive glazing | Energy generation | Control system

**Operation mode** Passive | Active

**Trigger** Thermal | Electromagnetic | Optical | Air quality | Occupants | Other

**Response input** Extrinsic | Intrinsic | Self-adjusting | Users' control

**Degree of adaptability** On/Off | Gradual

**TRL** 9

### Siemens Headquarters (2016)
Henning Larsen Architects, Munich (DE)

**Fig. 4.41** View of "The Wings," installation within the main court of the Siemens Headquarters. Source: https://commons.wikimedia.org/wiki/File:The_Wings,_Siemens_HQ_Munich,_April_2017.jpg by Martin Falbisoner (own work). CC BY-SA 4.0 (https://creativecommons.org/licenses/by-sa/4.0/deed.en) via Wikimedia commons

**Introduction** Designed to merge tradition with the future, Siemens' new global headquarters combines a brave architectural design with high-efficiency technologies, becoming one of the most sustainable European buildings: Its operation consumes as few resources as possible so as the materials that built it up are particularly environmentally compatible.

**The Smart Feature** The edifice is fully controlled by smart solutions from Siemens building technologies, thanks to about 30,000 data points that continuously analyze the building's behavior.

**The technologies** Smart building technologies control all the building services engineering. A central building management platform collects data and evaluates them continuously. The system controls heating, ventilation, and air conditioning systems as well as the glass louvers on the façade. Occupants, using single control modules, can adjust their own room patterns to meet individual needs. At the same time, the controls make employees aware of the responsible use of resources, informing them when energy is being consumed unnecessarily. To make this possible, a Green Building Monitor evaluates all the building's consumption and measurement values in real time and supplements these data with information such as details on current weather conditions.

**The Performances** Compared to the former building complex, the new building reduces the $CO_2$ footprint by nearly 90%. The use of primary energy is reduced by 88.5%. The amount of primary energy required is more than 50% below the level specified by the German Energy Saving Ordinance (EnEV). The building requires 90% less electric power than the old building complex did. Water consumption is reduced by 75%. The heating, too, requires significantly less energy, which saves the equivalent of 550,000 L of heating oil each year.

**Intended use** Residential | Service industry | Specialized

**Building state of development** Concept/Prototype | Real-scale edifice | Under construction

**Climate** Tropical | Dry/Desertic | Temperate | Continental | Polar

**Application scale** Element/Material | Component | Façade system

**Type** Insulation | Glazing | Shading | Opening systems | Energy harvesting/generation device | Other

**Purpose** Thermal comfort | Visual performance | Acoustic comfort | Energy management/generation | IAQ | Interaction and users' control | Other

**Adaptive function** Collect and storage | Control | Prevent/Reject | Allow interaction

**Classification** High-performance | Selective | Dynamic façade | Adaptive glazing | Energy generation | Control system

**Operation mode** Passive | Active

**Trigger** Thermal | Electromagnetic | Optical | Air quality | Occupants | Other

**Response input** Extrinsic | Intrinsic | Self-adjusting | Users' control

**Degree of adaptability** On/Off | Gradual

**TRL** 9

**King Abdullah Petroleum Studies and Research Centre (2017)**
Zaha Hadid Architects, Riyadh (ARS)

**Fig. 4.42** Detail of the upper envelope of the KAPSRC from the inside. Source: https://commons.wikimedia.org/wiki/File:DSC04791kapsarkrequest.jpg by Fahad Abdualhameid (own work) CC0 1.0 (https://creativecommons.org/publicdomain/zero/1.0/deed.en) via Wikimedia commons

**Introduction** KAPSARC is a nonprofit institution for independent research about policies that contribute to the most effective use of energy. The campus incorporates five buildings, conceived as a unified whole and designed in response to the environmental conditions of the surrounding to minimize energy and resource consumption according to a hexagonal prismatic honeycomb structure. The whole building can be seen as a metaphor of the traditional Arabic city or of the old cities, with small cells in the center that tend to form bigger and taller at the peripheries.

**The Smart Feature** Its design optimization, conceived as a living, organic structure, coupled with system selection allows obtaining a 45% reduction in energy performance when compared to the ASHRAE baseline standards.
The buildings, indeed, have been designed to minimize solar gain and are shaped to enclose shaded courtyards cooled by the prevailing desert winds. They are oriented to work naturally with the sun's path, providing both light and shade strategically at different times of day.
The particular self-supporting cellular structure provides flexible spaces that can be adapted to the multifunctional uses within the buildings.

**Intended use** Residential | Service industry | Specialized

**Building state of development** Concept/Prototype | Real-scale edifice | Under construction

**Climate** Tropical | Dry/Desertic | Temperate | Continental | Polar

**Application scale** Element/Material | Component | Façade system

**Type** Insulation | Glazing | Shading | Opening systems | Energy harvesting/generation device | Other

**Purpose** Thermal comfort | Visual performance | Acoustic comfort | Energy management/generation | IAQ | Interaction and users' control | Other

**Adaptive function** Collect and storage | Control | Prevent/Reject | Allow interaction

**Classification** High-performance | Selective | Dynamic façade | Adaptive glazing | Energy generation | Control system

**Operation mode** Passive | Active

**Trigger** Thermal | Electromagnetic | Optical | Air quality | Occupants | Other

**Response input** Extrinsic | Intrinsic | Self-adjusting | Users' control

**Degree of adaptability** On/Off | Gradual

**TRL** 9

**Metropole Rouen Normandy (2017)**
Ferrier Marchetti Studio, Rouen (FR)

**Fig. 4.43** View of the Metropole Rouen from the harbor. Source: https://commons.wikimedia.org/wiki/File:Hôtel_de_la_Métropole_Rouen_Normandie.jpg by Teddy76340 (own work). CC BY-SA 4.0 (https://creativecommons.org/licenses/by-sa/4.0/deed.en) via Wikimedia commons

**Introduction** Designed to fit in the surroundings, this particular building exploits the technological component to reach for a synthesis between the formal aspect and the performance aspect, able to foster its integration, thanks to its dynamic envelope able to offer even advantages in terms of performance.

**The Smart Feature** The dynamic building envelope, inspired by Monet's paintings of the Cathedral of Rouen, is able to confer the building a changing appearance throughout the day.

**The technologies** The building is equipped with double-layered façades whose external skin is made of individual glass elements, covered with a layer of metal oxides that give them changing chromatics, enlightened by the presence of water all around. On the roof, the glass scales make way for solar panels that provide a considerable contribution to the building's energy self-sufficiency.

**Intended use** Residential I Service industry I <u>Specialized</u>

**Building state of development** Concept/Prototype I <u>Real-scale edifice</u> I Under construction

**Climate** Tropical I Dry/Desertic I <u>Temperate</u> I Continental I Polar

**Application scale** Element/Material | Component | Façade system

**Type** Insulation | Glazing | Shading | Opening systems | Energy harvesting/generation device | Other

**Purpose** Thermal comfort | Visual performance | Acoustic comfort | Energy management/generation | IAQ | Interaction and users' control | Other

**Adaptive function** Collect and storage | Control | Prevent/Reject | Allow interaction

**Classification** High-performance | Selective | Dynamic façade | Adaptive glazing | Energy generation | Control system

**Operation mode** Passive | Active

**Trigger** Thermal | Electromagnetic | Optical | Air quality | Occupants | Other

**Response input** Extrinsic | Intrinsic | Self-adjusting | Users' control

**Degree of adaptability** On/Off | Gradual

**TRL** 8

**US Embassy (2017)**
Kieran Timberlake, London (UK)

**Fig. 4.44** View of the US Embassy building from the riverside. Source: https://commons.wikimedia.org/wiki/File:New_US_Embassy_Building,_Nine_Elms_(geograph_5628059).jpg by Paul Farmer (own work). CC BY-SA 2.0 (https://creativecommons.org/licenses/by-sa/2.0/deed.en) via Wikimedia commons

**Introduction** The United States embassy in Britain aims at combining high safety standards with an appearance of openness and transparency, making use of responsibly and sustainable materials to demonstrate the attention toward the environment.

**The Smart Feature** The laminated glazing façade, enveloped with a transparent film of ETFE, gives it a changeable translucent appearance.

**The technologies** The glazed envelope is wrapped on the most exposed sides with sails in ETFE that reflect the radiation, modifying their chromatics according to weather and sun position; even the particular form of the façade is designed to minimize solar gain and glare, allowing natural light to enter. The building integrates renewable energy sources, such as solar panels embedded in its envelope.

**Intended use** Residential | Service industry | Specialized

**Building state of development** Concept/Prototype | Real-scale edifice | Under construction

**Climate** Tropical | Dry/Desertic | Temperate | Continental | Polar

**Application scale** Element/Material | Component | Façade system

**Type** Insulation | Glazing | Shading | Opening systems | Energy harvesting/generation device | Other

**Purpose** Thermal comfort | Visual performance | Acoustic comfort | Energy management/generation | IAQ | Interaction and users' control | Other

**Adaptive function** Collect and storage | Control | Prevent/Reject | Allow interaction

**Classification** High-performance | Selective | Dynamic façade | Adaptive glazing | Energy generation | Control system

**Operation mode** Passive | Active

**Trigger** Thermal | Electromagnetic | Optical | Air quality | Occupants | Other

**Response input** Extrinsic | Intrinsic | Self-adjusting | Users' control

**Degree of adaptability** On/Off | Gradual

**TRL** 8

**The Bund Finance Centre (2017)**
Heatherwick and Foster + Partners, Shangai (RPC)

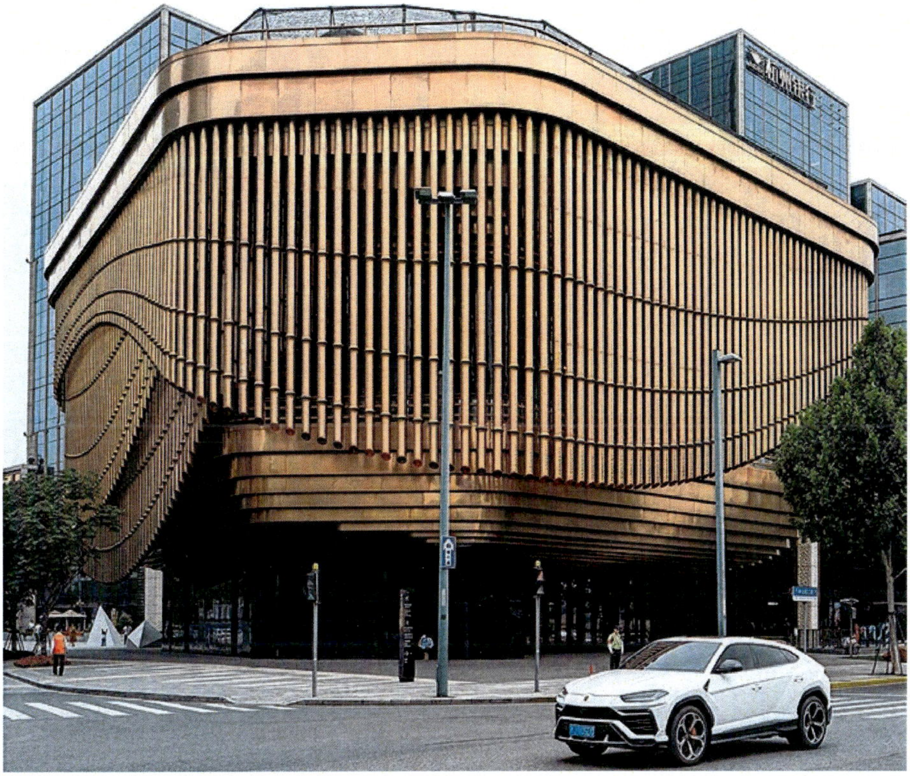

**Fig. 4.45** The Bund Finance Centre. Source: https://commons.wikimedia.org/wiki/File:The_Bund_Finance_Center_in_Shanghai.jpg by GillyBerlin via Flickr (https://www.flickr.com/photos/gillyberlin/48280189552/). CC BY 2.0 (https://creativecommons.org/licenses/by/2.0/deed.en) via Wikimedia commons

**Introduction** It is the major mixed-use building in Shanghai, born to redevelop its waterfront, acting as a connection between the old town and the new financial district. The Bund Finance Center is actually composed of eight buildings, for about 420,000 $m^2$ in which stand premium offices, a boutique hotel, a cultural center, and a wide variety of luxury retail spaces, all arranged around a landscaped public plaza.

**The Smart Feature** The cultural center is circled by a moving skin, which adapts to the changing use of the buildings and reveals the stages and views. It is made of stainless steel tubes inspired by the shape of bamboo poles, forming a flexible veil that can be adjusted to completely transform the appearance of the building.

**The technologies** The façade is organized with three moving layers that hang from the third floor and move along tracks powered by electric motors. It is made up of 675 individual magnesium alloy tassels, a reference to the traditional Chinese bridal headdress. The tassels range in length from around 2–16 m so that as each track independently moves, the veil rotates with the tassels overlapping and producing different visual effects and levels of opacity, reproducing the same way the sound of the wind through a bamboo forest.

**Intended use** Residential | Service industry | Specialized

**Building state of development** Concept/Prototype | Real-scale edifice | Under construction

**Climate** Tropical | Dry/Desertic | Temperate | Continental | Polar

**Application scale** Element/Material | Component | Façade system

**Type** Insulation | Glazing | Shading | Opening systems | Energy harvesting/generation device | Other

**Purpose** Thermal comfort | Visual performance | Acoustic comfort | Energy management/generation | IAQ | Interaction and users' control | Other

**Adaptive function** Collect and storage | Control | Prevent/Reject | Allow interaction

**Classification** High-performance | Selective | Dynamic façade | Adaptive glazing | Energy generation | Control system

**Operation mode** Passive | Active

**Trigger** Thermal | Electromagnetic | Optical | Air quality | Occupants | Other

**Response input** Extrinsic | Intrinsic | Self-adjusting | Users' control

**Degree of adaptability** On/Off | Gradual

**TRL** 8

**Futurium Belin (2017)**
Richter Musikowski, Berlin (DE)

**Fig. 4.46** The Futurium in Berlin. Source: https://commons.wikimedia.org/wiki/File:Futurium_Ostfassade.jpg by Boonekamp (own work). CC BY-SA 4.0 (https://creativecommons.org/licenses/by-sa/4.0/deed.en) via Wikimedia commons

**Introduction** Innovative building born to provide a forum for dialog on the future, in which all the people can actively participate. Conceived as a nearly zero energy building, it is organized into three levels, each tailored to a specific function.

**The Smart Feature** The modular kaleidoscopic glazed façade system, which combines translucent textured glass on the front with a folded metal reflector on the back, is the most striking feature of the Futurium. The optical interplay of reflective, translucent, and transparent surfaces results in smooth and shimmering skin, continuously transforming over the day depending on the lighting conditions and viewing angle. Moreover, the façade design combines such visual complexity with a modular ventilated system that allows seamless integration of openings.

**The technologies** The glazed envelope is made of more than 7500 units; this allowed the complete prefabrication and configuration of the diamond-shaped cassettes in the factory. Each cassette measures 700 mm × 700 mm and is composed of roughcast glazing printed with a white ceramic frit and affixed to folded stainless steel plates. It is the first Structural Sealant Glazing (SSG) system in Germany without mechanical restraints.

**Intended use** Residential | Service industry | Specialized

**Building state of development** Concept/Prototype | Real-scale edifice | Under construction

**Climate** Tropical | Dry/Desertic | Temperate | Continental | Polar

**Application scale** Element/Material | Component | Façade system

**Type** Insulation | Glazing | Shading | Opening systems | Energy harvesting/generation device | Other

**Purpose** Thermal comfort | Visual performance | Acoustic comfort | Energy management/generation | IAQ | Interaction and users' control | Other

**Adaptive function** Collect and storage | Control | Prevent/Reject | Allow interaction

**Classification** High-performance | Selective | Dynamic façade | Adaptive glazing | Energy generation | Control system

**Operation mode** Passive | Active

**Trigger** Thermal | Electromagnetic | Optical | Air quality | Occupants | Other

**Response input** Extrinsic | Intrinsic | Self-adjusting | Users' control

**Degree of adaptability** On/Off | Gradual

**TRL** 8

| | | GENERAL INFORMATION | | | FIRST LEVEL ANALISYS | | SECOND LEVEL ANALISYS | | | | | | | F.E. |
|---|---|---|---|---|---|---|---|---|---|---|---|---|---|---|
| **IMG** | **Year** | **Intended use** | **Bld. state of development** | **Climate condition** | **Application scale** | **Type** | **Purposes** | **Adaptive function** | **Classification** | **Operation mode** | **Trigger** | **Response imputs** | **Degree of adaptability** | **TRL** |
|  | 1991 | Residential | Concept / Protoype | Continental | Façade system | Energy harvesting / generation device | Energy manag./gen. | Prevent / Reject | High-performance | Passive | Optical | Extrinsic | Gradual | 6 |
|  | 1994 | Residential | Real-scale edifice | Temperate | Façade system | Other | Thermal comfort | Control | High-performance | Passive | Optical | Extrinsic | Gradual | 8 |
|  | 2002 | Specialized | Real-scale edifice | Temperate | Façade system | Opening systems | Interaction and users' control | Prevent / Reject | Dynamic façade | Active | Other | Users' control | Gradual | 9 |
|  | 2002 | Residential | Concept / Protoype | Temperate | Component | Shading | Visual performance | Prevent / Reject | Dynamic façade | Active | Occupants | Extrinsic | On/Off | 7 |

| | | GENERAL INFORMATION | | | FIRST LEVEL ANALISYS | | SECOND LEVEL ANALISYS | | | | | | | F.E. |
|---|---|---|---|---|---|---|---|---|---|---|---|---|---|---|
| IMG | Year | Intended use | Bld. state of development | Climate condition | Application scale | Type | Purposes | Adaptive function | Classification | Operation mode | Trigger | Response imputs | Degree of adaptability | TRL |
|  | 2002 | Specialized | Real-scale edifice | Tropical | Façade system | Shading | Visual performance | Control | Selective | Passive | Optical | Intrinsic | Gradual | 9 |
|  | 2003 | Specialized | Real-scale edifice | Temperate | Element / Material | Other | IAQ | Control | High-performance | Passive | Optical | Intrinsic | On/Off | 9 |
|  | 2003 | Specialized | Real-scale edifice | Temperate | Component | Energy harvesting / generation device | Energy manag./gen. | Prevent / Reject | Energy generation | Passive | Optical | Intrinsic | Gradual | 9 |
|  | 2004 | Service Industry | Real-scale edifice | Temperate | Component | Glazing | Visual performance | Prevent / Reject | Selective | Passive | Optical | Intrinsic | On/Off | 9 |

| IMG | Year | GENERAL INFORMATION | | | FIRST LEVEL ANALISYS | | SECOND LEVEL ANALISYS | | | | | | | F.E. |
|---|---|---|---|---|---|---|---|---|---|---|---|---|---|---|
| | | Intended use | Bld. state of development | Climate condition | Application scale | Type | Purposes | Adaptive function | Classification | Operation mode | Trigger | Response imputs | Degree of adaptability | TRL |
|  | 2004 | Service industry | Real-scale edifice | Tropical | Façade system | Other | Visual performance | Control | High-performance | Active | Electromagnetic | Extrinsic | On/Off | 8 |
|  | 2005 | Service Industry | Real-scale edifice | Temperate | Façade system | Other | Other | Control | High-performance | Active | Other | Users' control | On/Off | 9 |
|  | 2007 | Service Industry | Real-scale edifice | Temperate | Component | Shading | Visual performance | Control | Selective | Active | Optical | Users' control | Gradual | 9 |
|  | 2009 | Service Industry | Real-scale edifice | Temperate | Component | Shading | Visual performance | Control | Selective | Active | Other | Users' control | Gradual | 8 |

| | | GENERAL INFORMATION | | | FIRST LEVEL ANALISYS | | SECOND LEVEL ANALISYS | | | | | | | F.E. |
|---|---|---|---|---|---|---|---|---|---|---|---|---|---|---|
| IMG | Year | Intended use | Bld. state of development | Climate condition | Application scale | Type | Purposes | Adaptive function | Classification | Operation mode | Trigger | Response imputs | Degree of adaptability | TRL |
|  | 2009 | Residential | Concept / Protoype | Continental | Component | Other | Interaction and users' control | Allow interaction | Control system | Active | Occupants | Users' control | Gradual | 3 |
|  | 2009 | Specialized | Real-scale edifice | Tropical | Façade system | Glazing | IAQ | Control | High-performance | Passive | Other | Self-adjusting | Gradual | 9 |
|  | 2010 | Service Industry | Real-scale edifice | Continental | Component | Glazing | Thermal comfort | Control | High-performance | Passive | Optical | Intrinsic | Gradual | 9 |
|  | 2010 | Service Industry | Real-scale edifice | Temperate | Component | Shading | Visual performance | Control | Selective | Active | Occupants | Extrinsic | Gradual | 9 |

| | | GENERAL INFORMATION | | | FIRST LEVEL ANALISYS | | SECOND LEVEL ANALISYS | | | | | | | F.E. |
|---|---|---|---|---|---|---|---|---|---|---|---|---|---|---|
| IMG | Year | Intended use | Bld. state of development | Climate condition | Application scale | Type | Purposes | Adaptive function | Classification | Operation mode | Trigger | Response imputs | Degree of adaptability | TRL |
|  | 2010 | Service Industry | Real-scale edifice | Temperate | Component | Shading | Visual performance | Control | Selective | Passive | Optical | Intrinsic | Gradual | 9 |
|  | 2011 | Specialized | Concept / Protoype | Temperate | Component | Energy harvesting / generation device | Thermal comfort | Collect and storage | High-performance | Passive | Other | Extrinsic | Gradual | 6 |
|  | 2011 | Service Industry | Concept / Protoype | Temperate | Façade system | Other | Thermal comfort | Collect and storage | Control system | Passive | Optical | Users' control | Gradual | 7 |
| 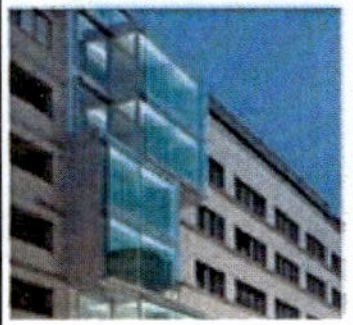 | 2011 | Service industry | Real-scale edifice | Continental | Component | Shading | Visual performance | Prevent / Reject | Selective | Passive | Other | Intrinsic | On/Off | 9 |

| | | GENERAL INFORMATION | | | FIRST LEVEL ANALISYS | | SECOND LEVEL ANALISYS | | | | | | | F.E. |
|---|---|---|---|---|---|---|---|---|---|---|---|---|---|---|
| IMG | Year | Intended use | Bld. state of development | Climate condition | Application scale | Type | Purposes | Adaptive function | Classification | Operation mode | Trigger | Response imputs | Degree of adaptability | TRL |
|  | 2011 | Specialized | Real-scale edifice | Continental | Component | Shading | Visual performance | Collect and storage | Dynamic façade | Active | Electromagnetic | Extrinsic | Gradual | 9 |
|  | 2011 | Service industry | Real-scale edifice | Continental | Component | Shading | Visual performance | Control | Selective | Passive | Optical | Extrinsic | Gradual | 9 |
|  | 2012 | Service Industry | Real-scale edifice | Temperate | Component | Shading | Visual performance | Control | Selective | Active | Optical | Extrinsic | Gradual | 8 |
|  | 2012 | Residential | Concept / Protoype | Temperate | Component | Energy harvesting / generation device | Thermal comfort | Collect and storage | High-performance | Passive | Optical | Extrinsic | On/Off | 7 |

| | | GENERAL INFORMATION | | | FIRST LEVEL ANALISYS | | SECOND LEVEL ANALISYS | | | | | | | F.E. |
|---|---|---|---|---|---|---|---|---|---|---|---|---|---|---|
| IMG | Year | Intended use | Bld. state of development | Climate condition | Application scale | Type | Purposes | Adaptive function | Classification | Operation mode | Trigger | Response imputs | Degree of adaptability | TRL |
|  | 2012 | Residential | Concept / Protoype | Temperate | Façade system | Energy harvesting / generation device | Energy manag./gen. | Collect and storage | Energy generation | Passive | Optical | Intrinsic | On/Off | 7 |
|  | 2012 | Specialized | Real-scale edifice | Temperate | Component | Shading | Visual performance | Prevent / Reject | Selective | Active | Optical | Intrinsic | Gradual | 9 |
|  | 2011 | Specialized | Real-scale edifice | Temperate | Element / Material | Other | Interaction and users' control | Allow interaction | Dynamic façade | Passive | Other | Intrinsic | Gradual | 9 |
|  | 2012 | Specialized | Real-scale edifice | Continental | Façade system | Opening systems | Visual performance | Control | Dynamic façade | Active | Electromagnetic | Intrinsic | Gradual | 8 |

| | | GENERAL INFORMATION | | | FIRST LEVEL ANALISYS | | SECOND LEVEL ANALISYS | | | | | | | F.E. |
|---|---|---|---|---|---|---|---|---|---|---|---|---|---|---|
| IMG | Year | Intended use | Bld. state of development | Climate condition | Application scale | Type | Purposes | Adaptive function | Classification | Operation mode | Trigger | Response imputs | Degree of adaptability | TRL |
|  | 2012 | Service Industry | Real-scale edifice | Dry/Desertic | Component | Shading | Visual performance | Control | Dynamic façade | Passive | Optical | Intrinsic | Gradual | 8 |
|  | 2012 | Service Industry | Real-scale edifice | Continental | Façade system | Shading | Energy manag./gen. | Control | Dynamic façade | Passive | Optical | Extrinsic | Gradual | 7 |
|  | 2013 | Residential | Concept / Protoype | Temperate | Component | Energy harvesting / generation device | Other | Collect and storage | High-performance | Passive | Electromagnetic | Intrinsic | Gradual | 7 |
| 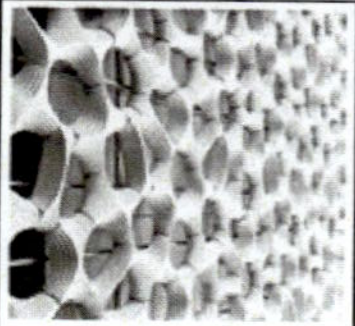 | 2013 | Specialized | Real-scale edifice | Temperate | Component | Other | IAQ | Prevent / Reject | High-performance | Passive | Optical | Intrinsic | On/Off | 8 |

| | | GENERAL INFORMATION | | | FIRST LEVEL ANALISYS | | SECOND LEVEL ANALISYS | | | | | | | F.E. |
|---|---|---|---|---|---|---|---|---|---|---|---|---|---|---|
| IMG | Year | Intended use | Bld. state of development | Climate condition | Application scale | Type | Purposes | Adaptive function | Classification | Operation mode | Trigger | Response imputs | Degree of adaptability | TRL |
|  | 2014 | Service Industry | Real-scale edifice | Continental | Component | Glazing | Visual performance | Control | Adaptive glazing | Active | Electromagnetic | Extrinsic | Gradual | 9 |
|  | 2014 | Specialized | Real-scale edifice | Continental | Element / Material | Shading | Visual performance | Control | Dynamic façade | Passive | Other | Intrinsic | Gradual | 9 |
|  | 2014 | Specialized | Real-scale edifice | Temperate | Component | Shading | Visual performance | Control | Selective | Active | Electromagnetic | Extrinsic | Gradual | 9 |
|  | 2014 | Residential | Real-scale edifice | Temperate | Component | Other | Interaction and users' control | Allow interaction | Control system | Active | Occupants | Users' control | Gradual | 8 |

| | | GENERAL INFORMATION | | | FIRST LEVEL ANALISYS | | SECOND LEVEL ANALISYS | | | | | | | F.E. |
|---|---|---|---|---|---|---|---|---|---|---|---|---|---|---|
| IMG | Year | Intended use | Bld. state of development | Climate condition | Application scale | Type | Purposes | Adaptive function | Classification | Operation mode | Trigger | Response imputs | Degree of adaptability | TRL |
|  | 2014 | Specialized | Concept / Protoype | Continental | Component | Shading | Visual performance | Control | Selective | Active | Occupants | Users' control | Gradual | 8 |
|  | 2015 | Specialized | Concept / Protoype | Temperate | Façade system | Insulation | Thermal comfort | Prevent / Reject | High-performance | Active | Other | Extrinsic | On/Off | 7 |
|  | 2015 | Specialized | Real-scale edifice | Tropical | Façade system | Other | Other | Allow interaction | High-performance | Passive | Other | Intrinsic | On/Off | 9 |
|  | 2016 | Service Industry | Real-scale edifice | Temperate | Component | Other | Thermal comfort | Prevent / Reject | High-performance | Passive | Other | Intrinsic | On/Off | 9 |

| | | GENERAL INFORMATION | | | FIRST LEVEL ANALISYS | | SECOND LEVEL ANALISYS | | | | | | | F.E. |
|---|---|---|---|---|---|---|---|---|---|---|---|---|---|---|
| IMG | Year | Intended use | Bld. state of development | Climate condition | Application scale | Type | Purposes | Adaptive function | Classification | Operation mode | Trigger | Response imputs | Degree of adaptability | TRL |
|  | 2016 | Service Industry | Real-scale edifice | Temperate | Façade system | Other | Interaction and users' control | Allow interaction | Control system | Active | Other | Users' control | On/Off | 9 |
|  | 2017 | Specialized | Real-scale edifice | Dry/Desertic | Façade system | Other | Thermal comfort | Control | High-performance | Passive | Other | Intrinsic | On/Off | 9 |
|  | 2017 | Specialized | Real-scale edifice | Temperate | Component | Glazing | Visual performance | Control | Selective | Passive | Optical | Intrinsic | On/Off | 8 |
|  | 2017 | Service Industry | Real-scale edifice | Temperate | Component | Shading | Visual performance | Prevent / Reject | Selective | Passive | Optical | Intrinsic | On/Off | 8 |

| | | GENERAL INFORMATION | | | FIRST LEVEL ANALISYS | | SECOND LEVEL ANALISYS | | | | | | | F.E. |
|---|---|---|---|---|---|---|---|---|---|---|---|---|---|---|
| IMG | Year | Intended use | Bld. state of development | Climate condition | Application scale | Type | Purposes | Adaptive function | Classification | Operation mode | Trigger | Response imputs | Degree of adaptability | TRL |
|  | 2017 | Service industry | Real-scale edifice | Tropical | Façade system | Other | Other | Allow interaction | Dynamic façade | Active | Electromagnetic | Extrinsic | On/Off | 8 |
|  | 2017 | Specialized | Real-scale edifice | Continental | Element / Material | Glazing | Other | Control | Adaptive glazing | Passive | Optical | Intrinsic | On/Off | 8 |

## 4.2 Lessons Learned: The Role of Transparent Building Components

The sample of the buildings selected and presented in the present chapter (see Sect. 4.1.3) has allowed to compile the precedent comprehensive synthetic tables, which map the specific parameters identified in relation to the different technological solutions and applications within the domain of Smart Building Envelopes.

This *supporting database* provides a systematic characterization of Smart Building Envelopes, outlining strategies adopted in such systems, recognizing common patterns, and identifying unexplored concepts to help further development of high-potential, innovative smart building technologies.

First of all, this review has allowed highlighting the significant role that transparent building components play in building energy efficiency and internal environmental comfort. As a matter of fact, the analysis conducted on the sample investigated revealed that the majority of the examples investigated pertain to the application of advanced innovative technologies to building components (56%); 35% concerns whole façades systems while only 9% is related to single element or materials. Among these percentage, the almost totality of technological innovations and solutions adopted addresses transparent components, being mostly related to glazing (13%) or shading (37%) systems; significant is also the percentage related to energy harvesting/generation devices (13%) as well systems related to other technologies (31%). Marginal is the amount of examples oriented toward opening systems (intended as those merely aimed at the movement of such components) and insulation, respectively, at 4% and 2%.

It follows that the main purpose addressed is the one related to visual performance, with a percentage of examples that fit within this category of 46%. Other relevant are thermal comfort (17%), interaction and users' control (11%), energy management/generation (9%), and IAQ (7%); thus, the main adaptive function is the one related to the control (50%), followed by the prevent/reject function (24%), the collect and storage function (13%), and the ability to allow users' interaction with the building system (13%).

From these considerations descends the classification given for the sample investigated, according to the (main) purposes for which the advanced technologies have been adopted. Such classification shows that most of the examples fit into the "high-performance" category (33%); the following are the "selective" category (30%), and the category of "dynamic façades" (20%), while secondary is the "control system" category (9%), followed at the end by the "adaptive glazing" category, and the "energy-generating" category (both with the 4%). Within the sample investigated, the greater majority of buildings presents a passive operation mode (59%) and a gradual degree of adaptability (61%), while the trigger of their adaptive features is most of optical nature (46%) even if occupants' presence (11%), electromagnetic (15%), and other kinds of trigger (28%) are relevant.

Final considerations about the type of response inputs have been made, showing that most of the technologies investigated through this analysis present an intrinsic

(46%) response; only 33% is activated through extrinsic inputs, while 19% operate after input of human nature.

At the end, regarding the TRL associated with technologies investigated, it emerges that the "smart" features of the sample investigated is greatly concerned with "*actual systems proven in the operational environment,*" thus means at TRL 9 (with a percentage of 50%), proving in this way the research hypothesis for which the "smartness," in the building environment, could be associated even with traditional materials and technologies, already seeped through current building practices, although innovated in their application and mutual behavior.

## References


100 11th Avenue by Jean Nouvel. (n.d.) Wiki Arquitectura. https://en.wikiarquitectura.com/building/100-11th-avenue-building/. Accessed 26 Nov 2020

Adaptive Solar Façade, (n.d.) project description. Available at: https://bta.climate-kic.org/stories/adaptive-solar-façade/. Accessed 09 Apr 2021

Adaptive Solar Façade (ASF), (n.d.) research page. Available at: https://systems.arch.ethz.ch/research/adaptive-solar-façade. Accessed 09 Apr 2021

Alenei L et al (2018) (ed), Case Studies - Adaptive Façades Network, TU Delft Open for The COST Action 1403 Adaptive Façade Network

Petersen Architekten (2011), Friedrichstrasse 40 Office Building / Petersen Architekten. Available via ArchDaily. https://www.archdaily.com/139547/friedrichstrasse-40-office-building-petersen-architekten Accessed 23 Feb 2021

Wingårdh Arkitektkontor (2012), Kuggen / Wingårdh Arkitektkontor. Available via ArchDaily. https://www.archdaily.com/289856/kuggen-wingardh-arkitektkontor. Accessed 23 Feb 2021

Arup (2013), The World's First Green Building Façades with Microalgae. Available at: https://www.youtube.com/watch?v=5r3p3NJVq_k

ARUP, (n.d.) Worldwide first façade system to cultivate micro-algae to generate heat and biomass as renewable energy sources. Available at: https://www.arup.com/projects/solar-leaf Accessed 14 Jan 2021

Atelier Jean Nouvel, (n.d.) 100 11th Avenue. http://www.jeannouvel.com/projets/100-11th-avenue/. Accessed 26 Nov 2020

ATIProject (2016), Forti Holding S.P.A. // Headquarters and office building, via Divisare.com. Available at: https://divisare.com/projects/307528-atiproject-forti-holding-s-pa-headquarters-and-office-building Accessed 18 Jan 2021

ATIProject (2017), Centro direzionale "Umberto Forti", via The Plan.it. Available at: https://www.theplan.it/award-2017-officebusiness/centro-direzionale-umberto-forti-1 Accessed 18 Jan 2021

ATIProject, (n.d.) Centro Direzionale "Umberto Forti". Available at: https://atiproject.com/project/umberto-forti-headquarters/ Accessed 18 Jan 2021

Woods Bagot, (n.d.) South Australian Health and Medical Research Institute. https://www.woodsbagot.com/projects/south-australian-health-and-medical-research-institute/ Accessed 21 Dec 2020

Ban S (2008), Museum of Paper Art. In: Jeska S, Transparent Plastics, Birkhäuser, Basel, pp 104-109. https://doi.org/10.1007/978-3-7643-8287-2_15

Basulto D (2010), US Embassy in London / KieranTimberlake Architects, via Archdaily.com. Available at: https://www.archdaily.com/50922/us-embassy-in-london-kieran-timberlake Accessed 18 Jan 2021

BI PV Project Details (n.d.) http://www.pvdatabase.org/projects_view_detailsmore.php?ID=256 Accessed 22 Feb 2021

Bishop E, Wilson J (2011), Concept to Realisation: GRC and Complex Façades. Paper presented at the 16th International Congress of the GRCA (GRC2011), Istanbul, September 2011.

Block I (2017), Kieran Timberlake completes £750 million "crystalline cube" for US Embassy in London, via Deezen.com. Available at: https://www.dezeen.com/2017/12/18/kieran-timberlake-us-embassy-london-nine-elms-american-architecture-uk/ Accessed 18 Jan 2021

Bojovic M (2013), Umbrella Mechanism for Madrid Pavillon. Via eVolo. Available at: https://www.evolo.us/umbrella-mechanism-for-madrid-pavilion/ Accessed 18 Jan 2021

Brake AG (2015), Henning Larsen's university building has a façade that moves in response to changing heat and light, via Deezen.com. Available at: https://www.dezeen.com/2015/07/14/henning-larsen-syddansk-universitet-sdu-kolding-campus-building-denmark-green-standards-university/ Accessed 15 Jan 2021

Breathing skins (2016), Breathing Skins Showroom Available at: https://vimeo.com/158980746

Brownell B, Swackhamer M (2015), Hypernatural. Architecture's New Relationship with Nature, Princeton Architectural Press

Bullivant L (2010), MEDIA TIC. ENRIC RUIZ GELI CLOUD 9. Available via theplan.it, https://www.theplan.it/eng/architettura/media-tic. Accessed 21 Dec 2020

Calatrava S (2011), Milwaukee Art Museum Takes Flight. Available via solaripedia. http://www.solaripedia.com/13/375/5106/milwaukee_art_museum_ground_floor_plan.html Accessed 23 Feb 2021

Cambiaso F (2015), UNSTUDIO: interactive and dynamic façades in Seoul. Available via Alchimag.net https://alchimag.net/en/architecture/unstudio-interactive-and-dynamic-façades-in-seoul/ Accessed 21 Dec 2020

Caniggia G, Maffei GL (1979), Composizione architettonica e tipologia edilizia, Marsilio, Venezia

Căplescu OA (2013) Augmented Spaces. A Look Beyond Advertising. Urbanism. Arhitectură. Construcţii, 4(3):1-17

Casini M (2016), Smart Buildings. Advanced Materials and Nanotechnology to Improve Energy-Efficiency and Environmental Performance, Woodhead Publishing Series in Civil and Structural Engineering: Number 69, Elsevier.

Chaix & Morel et associés, JSWD Architekten (2013), Q1, ThyssenKrupp Quarter Essen / JSWD Architekten + Chaix & Morel et Associés. Available via Archdaily.com, https://www.archdaily.com/326747/q1-thyssenkrupp-quarter-essen-jswd-architekten-chaix-morelet-associes. Accessed 17 Dec 2020

Chalcraft E (2012), Windswept by Charles Sowers. Available via Deezen.com https://www.dezeen.com/2012/11/06/windswept-installation-by-charles-sowers/ Accessed 21 Dec 2020

Jubilee Church (2003), Richard Meier & Partners Architects LLP. http://www.richardmeier.com/?projects=jubilee-church-2. Accessed 24 Nov 2020

Cilento K (2010), 100 Eleventh Avenue / Jean Nouvel. ArchDaily.com. https://www.archdaily.com/76113/100-eleventh-avenue-jean-nouvel. Accessed 26 Nov 2020

Cilento K (2012), Al Bahar Towers Responsive Façade / Aedas. Available via ArchDaily.com https://www.archdaily.com/270592/al-bahar-towers-responsive-façade-aedas Accessed 21 Dec 2020

CLOUD-9 - Enric Ruiz-Geli (2012), MEDIA-ICT. Available via Divisare.com, https://divisare.com/projects/194744-cloud-9-enric-ruiz-geli-media-ict. Accessed 21 Dec 2020

Conforti C, Marandola M (2009), Richard Meier. Minimum. Biblioteca essenziale di architettura. Motta Editore

Cornachio J (2017), Behind the Building: Richter Musikowski's Fantastic Futurium. Available via Architizer.com. https://architizer.com/blog/practice/materials/behind-building-richter-musikowski-futurium/ Accessed 25 Feb 2021

Danisi E (2017), Achieving an uncompromising architectural aesthetic with ambitious environmental credentials. Available via ARUP.com. https://www.arup.com/projects/kapsarc Accessed 25 Feb 2021

Denz PR, Priedemann W, Anders L (2018), ACT Façade – Interior sun shading for energy efficient fully glazed façades. Paper presented at the final conference of COST TU1403 "Adaptive Façades Network" FAÇADE 2018, Lucerne, 26-27 November 2018

Despland-Lichtert N (2017), Take a look at the kinetic façade of Foster + Partners and Heatherwick Studio's new Bund Finance Centre in Shanghai. Available via Archinect.com. https://archinect.com/news/article/150033188/take-a-look-at-the-kinetic-façade-of-foster-partners-and-heatherwick-studio-s-new-bund-finance-centre-in-shanghai Accessed 25 Feb 2021

Detail daily (2013), Fighting foulness: de-polluting façade by Elegant Embellishments, via Detail-online. Available at: https://www.detail-online.com/article/fighting-foulness-de-polluting-façade-by-elegant-embellishments-24764/ Accessed 14 Jan 2021

Doroteo J (2016), Let Your Building "Breathe" With This Pneumatic Façade Technology, via Archdaily.com. Available at: https://www.archdaily.com/789230/let-your-building-to-breathe-with-this-pneumatic-façade-technology Accessed 18 Jan 2021

Dragone R (2010), Milano inaugura il Palazzo che respira. Available via ArchiPortale.com, http://www.archiportale.com/news/2010/03/architetura/milanoinaugura-il-palazzo-che-respira_17995_3.html. Accessed 26 Nov 2020

Dunay, Wheeler + Schubert (2013), Lumenhaus. In: Trubiano F (ed), Design and Construction of High performance homes. Building Envelopes, Renewable Energies and Integrated Practice, Routledge

Econtrol-glas, (n.d.) FESTO Automation Center in Esslingen. Available at: https://www.ec-smartglass.com/en/company/references/show/4 Accessed 14 Jan 2021

Elegant embellishments, (n.d.) Decorative façades that eat pollution. Available at: http://www.elegantembellishments.net/home-1/ Accessed 14 Jan 2021

Esplanade – Theatres on the Bay. (n.d.) Available via DPA.com. https://dpa.com.sg/projects/esplanadetheatresonthebay/ Accessed 19 Feb 2021

European Commission (2019), HORIZON 2020 – WORK PROGRAMME 2018-2020 General Annexes, Extract from Part 19 - Commission Decision C(2017)7124[3]. https://ec.europa.eu/research/participants/data/ref/h2020/other/wp/2018-2020/annexes/h2020-wp1820-annex-ga_en.pdf Accessed 1 Feb 2021

Fearson A (2014), UNStudio's dynamic façade will control indoor climate for Seoul tower. Available via Deezen.com https://www.dezeen.com/2014/04/30/hanwha-hq-seoul-unstudio-dynamic-façade/ Accessed 21 Dec 2020

Fernandez Solla I (2010), ThyssenKrupp Quarter façades: a giant's gentle skin. Available via Façades Confidential. http://façadesconfidential.blogspot.com/2010/12/thyssenkrupp-quarter-façades-giants.html Accessed 23 Feb 2021

Ferrier Marchetti Studio (2017), Metropole Rouen Normandy, via Divisare.com. Available at: https://divisare.com/projects/370209-jacques-ferrier-architectureluc-boegly-metropole-rouen-normandy Accessed 18 Jan 2021

Ferrier Marchetti Studio, Métropole Rouen Normandie Headquarters, France, 2018. A carbon zero public building, on the banks of the Seine in Rouen. Available at: https://ferriermarchetti.studio/project/metropole-rouen-normandie Accessed 18 Jan 2021

Festo AG & Co KG, (n.d.) Festo Automation Center. Available at: https://www.festo.com/net/SupportPortal/Files/380783/Festo_AutomationCenter_EN.pdf Accessed 14 Jan 2021

Fisher T (2016), Shanghai Natural History Museum. Available via Architect magazine. https://www.architectmagazine.com/project-gallery/shanghai-natural-history-museum_o Accessed 25 Feb 2021

FJMT (2010), Surry Hills Library and Community Centre / FJMT. Available via ArchDaily. https://www.archdaily.com/57339/surry-hills-library-and-community-centre-fjmt Accessed 23 Feb 2021

Foster + Partners + Heatherwick Studio (2017), Bund Finance Centre / Foster + Partners + Heatherwick Studio. Available via ArchDaily. https://www.archdaily.com/881511/bund-finance-centre-foster-plus-partners-plus-heatherwick-studio Accessed 25 Feb 2021

Frearson A (2013), Umbrella Façade for the Madrid Pavillon by 3Gatti Architecture Studio, via Deezen.com. Available at: https://www.dezeen.com/2013/04/23/umbrella-façade-madrid-pavilion-by-3gatti/ Accessed 18 Jan 2021

Frearson A (2015), UNStudio completes a digitally controlled home with a flower-shaped plan, via Deezen.com. Available at: https://www.dezeen.com/2015/02/03/unstudio-digitally-controlled-home-wind-house-flower-shaped-plan-netherlands/ Accessed 15 Jan 2021

Furuto A (2012), Windswept Installation / Charlse Sowets Studios. Available via ArchDaily.com https://www.archdaily.com/214862/windswept-installation-charles-sowers-studios Accessed 21 Dec 2020

Gemini House (n.d.-a), available via Architectuul.com. http://architectuul.com/architecture/gemini-house. Accessed 19 Feb 2021

Gemini House (n.d.-b), available via Wikipedia, the free encyclopedia. https://en.wikipedia.org/wiki/Gemini_house Accessed 19 Feb 2021

Gerfen K (2010), ThyssenKrupp Quarter. Available via architectmagazine.com. https://www.architectmagazine.com/design/buildings/thyssenkrupp-quarter_o Accessed 17 Dec 2020

Giselbrecht E, (n.d.) Dynamic Façade (Kiefer Technic Showroom). Available via ArchiTonic.com. https://www.architonic.com/en/project/ernst-giselbrecht-partner-dynamic-façade-kiefer-technicshowroom/5100449. Accessed 25 Nov 2020

Gonchar J (2010), Hoberman's "Transformable Design" Gaining Momentum. Architectural record. https://www.architecturalrecord.com/articles/5374-hoberman-s-transformable-design-gainingmomentum. Accessed 25 Nov 2020

Gozdanic L (2016), Festo's new automated glass façade outsmarts Germany's strict energy standards, via inhabitat.com. Available at: https://inhabitat.com/festos-new-automated-glass-façade-outsmarts-germanys-strict-energy-standards/ Accessed 14 Jan 2021

Grisi T (2017), Facciata dell'Ospedale Eskenazi a Indianapolis, Rob Ley, via Arketipomakazine.it. Available at: https://www.arketipomagazine.it/facciata-dellospedale-eskenazi-a-indianapolis-rob-ley/ Accessed 15 Jan 2021

Grozdanic L (2016), Renovated Siemens Headquarters in Munich now consumes 90% less energy and 75% less water, via Inhabitat.com. Available at: https://inhabitat.com/renovated-siemens-headquarters-in-munich-now-consumes-90-less-energy-and-75-less-water/ Accessed 18 Jan 2021

Hamilton B, (n.d.) Al Bahr Towers. https://www.ahr.co.uk/Al-Bahr-Towers Accessed 21 Dec 2020

Heliotrop. (n.d.) Rotatable Solar House. Available via archilovers.com. https://www.archilovers.com/projects/44203/heliotrop.html#info Accessed 19 Feb 2021

Heliotrope (n.d.) (building) available via Wikipedia, the free encyclopedia. https://en.wikipedia.org/wiki/Heliotrope_(building) Accessed 19 Feb 2021

Henning Larsen Architects (2015a), Kolding Campus of the University of Southern Denmark, via Divisare.com. Available at: https://divisare.com/projects/301561-henning-larsen-architects-hufton-crow-kolding-campusof-the-university-of-southern-denmark Accessed 15 Jan 2021

Henning Larsen Architects (2015b), SDU Campus Kolding / Henning Larsen Architects, via Archdaily.com. Available at: https://www.archdaily.com/590576/sdu-campus-kolding-henning-larsen-architects Accessed 15 Jan 2021

Henning Larsen Architects (2015c), University of Southern Denmark – Campus Kolding. Available at: https://henninglarsen.com/en/projects/featured/0942-sdu-campus-kolding Accessed 15 Jan 2021

Henning Larsen Architects (2016), Siemens Headquarters / Henning Larsen Architects, via Archdaily.com. Available at: https://www.archdaily.com/791741/siemens-headquarters-henning-larsen-architects Accessed 18 Jan 2021

Henning Larsen Architects, (n.d.) Siemens' Global Headquarters. Available at: https://henninglarsen.com/en/projects/1100-1199/1133-siemens-global-headquarters Accessed 18 Jan 2021

IBA Hamburg GmbH (2012a). Smart is green. https://www.internationale-bauausstellung-hamburg.de/en/projects/the-building-exhibition-within-the-building-exhibition/smart-material-houses/smart-is-green/projekt/smart-is-green.html. Accessed 21 Dec 2020

IBA Hamburg GmbH (2012b). Soft house. https://www.internationale-bauausstellung-hamburg.de/en/projects/the-building-exhibition-within-the-building-exhibition/smart-material-houses/soft-house/projekt/soft-house.html. Accessed 21 Dec 2020

Inhabitat (2007), Extreme Solar Living: The Gemini Haus. Inhabitat.com. https://inhabitat.com/extreme-solar-the-gemini-haus/ Accessed 19 Feb 2021

International Building Exhibition IBA Hamburg, BIQ. (n.d.) Available at: https://www.internationale-bauausstellung-hamburg.de/en/themes-projects/the-building-exhibition-within-the-building-exhibition/smart-material-houses/biq/projekt/biq.html Accessed 14 Jan 2021

Jaafar AA, Halil ZA (2019), Kinetic Façade as a Tool For Energy Efficiency. International Journal of Engineering Research and Reviews, 7(4):1-7

Jacques Ferrier Architecture (2017), Headquarters of Métropole Rouen Normandie / Jacques Ferrier Architecture, via Archdaily.com. Available at: https://www.archdaily.com/882635/headquarters-of-metropole-rouen-normandie-jacques-ferrier-architecture Accessed 18 Jan 2021

Jayathissa et al. (2016), Energy Performance of PV Modules as Adaptive Building Shading Systems. Paper presented at the 32nd European Photovoltaic Solar Energy Conference and Exhibition, Munich, Germany, 20-24 June2016

Jayathissa et al. (2017a), Optimising building net energy demand with dynamic BIPV shading, Applied Energy 202:726-735

Jayathissa et al. (2017b), Sensitivity of Building Properties and Use Types for the Application of Adaptive Photovoltaic Shading Systems, Energy Procedia 122:139-144

Jordana S (2013), In Progress: SAHMRI / Woods Bagot / Woods Bagot. Available via ArchDaily.com https://www.archdaily.com/454746/in-progress-sahmriwoods-bagot-woods-bagot. Accessed 21 Dec 2020

JSWD Architekten, (n.d.) ThyssenKrupp Quarter. Available via world-architects.com. https://www.world-architects.com/en/projects/view/thyssenkrupp-quarter-1. Accessed 17 Dec 2020

Kido EM, Cywinski Z (2013) The new steel-glass architecture of buildings in Japan. Steel Construction 6(3):229-237. DOI: https://doi.org/10.1002/stco.201320012

Konstantinidis CV, (n.d.) Analysis Tools for High Performance Design. Available via School of Architecture, The University of Texas at Austin. https://soa.utexas.edu/sites/default/disk/analytic_tools/analytic_tools/09_03_fa_kienzl_konstantinidis_ml.pdf Accessed 19 Feb 2021

Kuggen. (n.d.-a) Available via Archello. https://archello.com/project/kuggen Accessed 23 Feb 2021

Kuggen. (n.d.-b) Available via Wikiarquitectura. https://en.wikiarquitectura.com/building/kuggen/ Accessed 23 Feb 2021

Kuggen. (n.d.-c) Wikipedia. https://en.wikipedia.org/wiki/Kuggen Accessed 23 Feb 2021

La Defense building. (n.d.). Available via Wikiarquitectura. https://en.wikiarquitectura.com/building/la-defense-building/ Accessed 25 Nov 2020

Laylin T (2011), Henning Larsen Wins Bid To Re-Structure Siemens Munich Headquarters, via Inhabitat.com. Available at: https://inhabitat.com/henning-larsen-wins-bid-to-re-structure-siemens-munich-headquarters/ Accessed 18 Jan 2021

Lomholt I (2016), Kiefer Technic Showroom. Available via e-architect.co.uk. https://www.e--architect.co.uk/austria/kiefer-technic-showroom. Accessed 25 Nov 2020

Loonen RCGM (2010) CABS – What can we simulate? Dissertation, University of Eindhoven

Lumenhaus. (n.d.) https://lumenhaus.com/index2.html. Accessed 24 Nov 2020

Maier F (2012), One Ocean – Thematic pavilion for EXPO 2012. Detail-online. https://www.detail-online.com/article/one-ocean-thematic-pavilion-for-expo-2012-16339/ Accessed 24 Feb 2021

Mairs J (2017), Zaha Hadid Architects reveals honeycomb-like oil research centre in Riyadh, via Deezen.com. Available at: https://www.dezeen.com/2017/10/26/zaha-hadid-architects-king-abdullah-petroleum-studies-research-centre-riyadh-saudi-arabia/ Accessed 18 Jan 2021

Richard Meier & Partners Architects (2009), Church of 2000. Available via ArchDaily.com. https://www.archdaily.com/20105/church-of-2000-richard-meier. Accessed 24 Nov 2020

Meinhold B (2010), Australia's Surry Hills Library Sets New Standard for Green Design. Available via Inhabitat https://inhabitat.com/surry-hills-library-australias-new-standard-of-sustainable-excellence/ Accessed 23 Feb 2021

Meinhold B (2011), Berlin Office Building Features Two Different High-Performance Façades. Available via Inhabitat. https://inhabitat.com/berlin-office-building-features-two-different-high-performance-façades/ Accessed 23 Feb 2021

Meinhold B (2012), Windswept is a Fascinating Kinetic Façade That Reveals the Direction of the Wind. Available via InHabitat.com https://inhabitat.com/windswept-is-a-fascinating-kinetic-façade-that-reveals-the-direction-of-the-wind/ Accessed 21 Dec 2020

Milwaukee Art Museum. (n.d.) The Quadracci Pavillion. https://mam.org/info/quadracci.php Accessed 23 Feb 2021

Nagy Z et al. (2016), The Adaptive Solar Façade: From Concept to Prototypes, Frontiers of Architectural Research. 5:143–156. https://doi.org/10.1016/j.foar.2016.03.002

Naidoo R (2010), enric ruiz-geli: media ICT building under construction in Barcelona. Available via designboom.com, https://www.designboom.com/architecture/enric-ruiz-geli-media-ict-building-under-construction/. Accessed 21 Dec 2020

North House on RVTR website. (n.d.) http://rvtr.com/projects/project-a/. Accessed 26 Nov 2020

Offices "La Defense". (n.d.) Available via german-architects.com. https://www.german-architects.com/fa/projects/view/offices-la-defense. Accessed 25 Nov 2020

Parthenopoulou NK and Malindretos M (2016), Materials Today: Proceedings 3, pp. 898-912

Pearman H (2017), U.S. Embassy in London by KieranTimberlake, via Architecturalrecord.com. Available at: https://www.architecturalrecord.com/articles/13162-us-embassy-in-london-by-kierantimberlake Accessed 18 Jan 2021

Peregoy B (2015), The New Perkins+Will–designed Shanghai Natural History Museum Is Open. Available via Architectural Digest. https://www.architecturaldigest.com/story/shanghai-natural-history-museum-perkins-and-will Accessed 25 Feb 2021

Perkins+Will (2015), Shanghai Natural History Museum / Perkins+Will. Available via ArchDaily. https://www.archdaily.com/623197/shanghai-natural-history-museum-perkins-will Accessed 25 Feb 2021

Petty M (2008) Chanel Ginza: The past, present and future of media skins. Professional Lighting Design, 60:21-29

prosolve307e, (n.d.) A pollution-eating façade module. Available at: http://www.prosolve370e.com Accessed 14 Jan 2021

Puglisi V (2015), Sistemi, tecnologie e materiali innovativi per gli involucri contemporanei. Aracne Editrice, Ariccia (RM).

Q1, ThyssenKrupp Quartier Essen. (n.d.) Available via Architonic.com, https://www.architonic.com/en/project/jswdarchitektenq1-thyssenkrupp-quartier-essen/5100714. Accessed 17 Dec 2020

Radwan AH (2016), Smart Façades. From adaptation with climate, changing esthetical values, till digital drawings on buildings envelope. https://d1wqtxts1xzle7.cloudfront.net/41020030/03_SMART_FaçadeS-Copy.pdf?1452353964=&response-content-disposition=inline%3B+filename%3DSmart_Façades_From_adaptation_with_clima.pdf&Expires=1614098945&Signature=TDD5cveV~uLRq0GWdKXwa3Vidu5Ur4fSm8rjJftPjJC79OmIXcBYzFesyxBOYfARxWkDwegfC3A9~KJhxowioaN2P1DPYG6aev5u8sR78H9KhuK0gEzUSc8OKvREK0TfDW7SmAa9gtdTU5-HXQYPlSGebJ22G-kvuMeRUQt51I25Gajs23aokXWhsxROM~bkCKBn4G45cagO18J~kek84M4thY-zLZtjMyY3sGiW8tdGKuOpzyhg1H6F6LraNNTRB5VtQ~wUkLu3Dd69LqW0V22W73kssfkPTwv4cKX4jNNRz64tKAU0dY5LAjaZak5nIakkJ4ZpjaAsh8B8VVoRpw__&Key-Pair-Id=APKAJLOHF5GGSLRBV4ZA Accessed 23 Feb 2021

Rhodes M (2014), A Massive Installation That Makes This Building Change Colors, via Wired.com. Available at: https://www.wired.com/2014/08/a-massive-installation-that-makes-this-building-change-colors/ Accessed 15 Jan 2021

Rinaldi M (2017), Headquarters of Métropole Rouen Normandie by Jacques Ferrier Architecture. Available at: https://aasarchitecture.com/2017/10/headquarters-metropole-rouen-normandie-jacques-ferrier-architecture.html/ Accessed 18 Jan 2021

Risen C (2014), South Australian Health and Medical Research Institute, Designed by Woods Bagot. Available via architectmagazine.com https://www.architectmagazine.com/design/buildings/south-australian-health-and-medical-research-institute-designed-by-woods-bagot_o Accessed 21 Dec 2020

(2013) RMIT Design Hub / Sean Godsell. Available via ArchDaily. https://www.archdaily.com/335620/rmit-design-hub-sean-godsell. Accessed 21 Dec 2020

Rob Ley Studio (2014a), 100 Parking Structure Art Façade / Rob Ley Studio, via Archdaily.com. Available at: https://www.archdaily.com/536756/parking-structure-art-façade-urbana Accessed 15 Jan 2021

Rob Ley Studio (2014b), May/September. Available at: https://www.rob-ley.com/portfolio?pgid=jk33kr7i-9e8dbfca-e0d8-4e93-b786-b03e785dbdf5 Accessed 15 Jan 2021

Rohlffs K, Gerber A, Kamps G and Disch R (1996), Energy characterization of the rotatable solar house HELIOTROP trademark in Freiburg, Germany. First results. Paper presented at the International Solar Forum (EuroSun '96) and at the 10 International Solar Forum of Deutsche Gesellschaft fuer Sonnenenergie e.V. (DGS), Freiburg im Breisgau (Germany), 16-19 Sep 1996

Rosenfield K (2014), UNStudio's Responsive Façade to Transform Seoul Office Tower. Available via ArchDaily.com https://www.archdaily.com/500006/unstudio-s-responsive-façade-concept-selected-to-transform-seoul-office-tower Accessed 21 Dec 2020

Enric Ruiz-Geli (2010), MEDIA-TIC. Available via Archdaily.com, https://www.archdaily.com/49150/media-tic-enric-ruiz-geli. Accessed 21 Dec 2020

Enric Ruiz-Geli, (n.d.) MEDIA-ICT BUILDING CFZB, 22@, BARCELONA. https://www.ruiz-geli.com/projects/built/media-tic.php. Accessed 21 Dec 2020

Scalco C (2017), Headquarter Forti Holding a Montacchiello (PI) – ATIProject, via Arketipo Magazine. Available at: https://www.arketipomagazine.it/headquarter-forti-holding-a-montacchiello-pi-atiproject/ Accessed 18 Jan 2021

Schoof J (2018), Iridescent lozenges made of cast glass: the Futurium in Berlin. Available via Detail-online. https://www.detail-online.com/article/iridescent-lozenges-made-of-cast-glass-the-futurium-in-berlin-32050/ Accessed 25 Feb 2021

Godsell Sean, (n.d.) RMIT Design Hub, https://www.seangodsell.com/rmit-design-hub. Accessed 21 Dec 2020

(2013), Una pelle reattiva e sostenibile: Centro di Design RMIT a Melbourne, Sean Godsell Architects. Available via architetti.com https://www.architetti.com/una-pelle-reattiva-e-sostenibile-centro-di-design-rmit-a-melbourne-sean-godsell-architects.html. Accessed 21 Dec 2020

Selimex Headoffices, (n.d.) Available via archello.com. https://archello.com/project/selimex-headoffices. Accessed 25 Nov 2020

Seng A (2010), Pola Ginza Building Façade. https://arthurtseng.wordpress.com/2010/01/13/pola-ginza-building-façade/. Accessed 26 Nov 2020

Sensual City Studio (2018), A history of Thresholds, Berlin: Jovis

Shanghai Natural History Museum. (n.d.) Perkins&Will. https://perkinswill.com/project/shanghai-natural-history-museum/ Accessed 25 Feb 2021

Sheikh WT, Asghar Q (2019) Adaptive biomimetic façades: Enhancing energy efficiency of highly glazed buildings. Frontiers of Architectural Research, 8(3):319-331.https://doi.org/10.1016/j.foar.2019.06.001.

Shigeru Ban Architects (2002), Paper Art Museum A, B – Mishima, Shizuoka, Japan. http://www.shigerubanarchitects.com/works/2002_paper-art-museum/index.html. Accessed 16 Nov 2020

Shigeru Ban Architects, (n.d.) Paper Art Museum. In: world-architects.com. Profiles of Selected Architects. https://www.world-architects.com/en/shigeru-ban-architects-tokyo/project/paper-art-museum. Accessed 16 Nov 2020

Siemens (2016), New corporate Headquarters, Munich. Available at: https://press.siemens.com/global/en/feature/new-corporate-headquarters-munich Accessed 18 Jan 2021

SOMA (2012), One Ocean, Thematic Pavilion EXPO 2012 / soma. Available via ArchDaily. https://www.archdaily.com/236979/one-ocean-thematic-pavilion-expo-2012-soma Accessed 24 Feb 2021

Charles Sowers, (n.d.) Windswept https://www.charlessowers.com/windswept Accessed 21 Dec 2020

Stoughton J (2016), Festo's AutomationCenter outsmarts German energy standards with a high performance automated façade, via The Architect's Newspaper. Available at: https://www.archpaper.com/2016/05/festos-automationcenter-outsmarts-german-energy-standards-high-performance-automated-façade/ Accessed 14 Jan 2021

Sullivan CC (2007) Changel Ginza, Tokio. Available via ArchitectMagazine.com. https://www.architectmagazine.com/technology/lighting/chanel-ginza-tokyo_o Accessed 22 Feb 2021

Surry Hills Library and community centre upgrade. (n.d.) City of Sydney. https://www.cityofsydney.nsw.gov.au/building-facility-upgrades/surry-hills-library-and-community-centre-upgrade Accessed 23 Feb 2021

Svetozarevic et al. (2016), SoRo-Track: A Two-Axis Soft Robotic Platform for Solar Tracking and Building-Integrated Photovoltaic Applications, Conference Proceedings 2016 IEEE International Conference on Robotics and Automation (ICRA), Stockholm, Sweden, (2016): 4945 - 4950

Svetozarevic B et al. (2019), Dynamic photovoltaic building envelopes for adaptive energy and comfort management, Nature Energy. (2019) 1. https://doi.org/10.1038/s41560-019-0424-0

The Esplanade – Theatres on the Bay. (n.d.) Available via Aterlierten.com. https://www.atelierten.com/projects/the-esplanade-theatres-on-the-bay/ Accessed 19 Feb 2021

The European Portal For Energy Efficiency in Buildings (2015), The BIQ House: first algae-powered building in the world. Available at: http://www.buildup.eu/en/node/43910 Accessed 14 Jan 2021

The Heliotrope. (n.d.) Available via roldisch.de. http://www.rolfdisch.de/en/projects/das-heliotrop-2/ Accessed 19 Feb 2021

Thün, Velikov (2013), North House: Climate Responsive Envelope and Controls Systems. In: Trubiano F (ed) Design and Construction of High performance homes. Building Envelopes, Renewable Energies and Integrated Practice, Routledge.

Kieran Timberlake, (n.d.) US Embassy in London. How can we build an embassy that reflects the core values of democracy—transparency, openness, and equality—and is welcoming, secure, and highly sustainable? Available at: https://kierantimberlake.com/pages/view/88/embassy-ofthe-united-states-of-america/parent:3 Accessed 18 Jan 2021

Tobe i&a, (n.d.) Breathing skins project. Available at: https://www.tebe.berlin/breathing-skins/ Accessed 18 Jan 2021

Transsolar Energietechnik GmbH. One Ocean – Pavillion EXPO 2012, Yeosu, South Korea. https://transsolar.com/projects/one-ocean-pavillon-expo-2012 Accessed 24 Feb 2021

Werner Tscholl (2005), Uffici Selimex. Available via Divisare.com. https://divisare.com/projects/124973-werner-tscholl-alexa-rainer-uffici-selimex. Accessed 25 Nov 2020

Werner Tschozll (2005), Selimex. http://www.werner-tscholl.com/new-constructions/selimex-latsch/. Accessed 25 Nov 2020

UNStudio (2003), Offices "La Defense". Available via Divisare.com. https://divisare.com/projects/143011-unstudio-christian-richters-offices-ladefense Accessed 25 Nov 2020

UNStudio (2015a), The W.I.N.D. House / UNStudio, via Arch.daily.com Available at: https://www.archdaily.com/595850/the-w-i-n-d-house-unstudio Accessed 15 Jan 2021

UNStudio (2015b), The W.I.N.D. House, via world-architects.com Available at: https://www.world-architects.com/ca/architecture-news/works/the-wind-house Accessed 15 Jan 2021

UNStudio (2015c), The W.I.N.D. House. Available at: https://www.unstudio.com/en/page/11968 Accessed 15 Jan 2021

UNStudio, (n.d.-a) "La Defense" Offices. https://www.unstudio.com/en/page/3289/ladefense-offices. Accessed 25 Nov 2020

UNStudio, (n.d.-b) Hanwha Headquarters Remodelling. https://www.unstudio.com/en/page/11994 Accessed 21 Dec 2020

Velikov, Bartram (2009) North House: Developing Intelligent Building technology and User Interface in Energy Independent domestic Environments. In: Architecture, Energy and the Occupant's Perspective: Proceedings of the PLEAS2009 - 26th Conference on Passive and Low Energy Architecture, Quebec City, Canada, 22-24 June 2009.

Vinnitskaya I (2010), Kiefer Technic Showroom / Ernst Giselbrecht + Partner. Available via ArchDaily.com. https://www.archdaily.com/89270/kiefer-tecnicshowroom-ernst-giselbrecht-partner. Accessed 25 Nov 2020

(2012) Virginia Tech LumenHAUS: A multifaceted, multi-tasking solar star. In: Vtech News. https://vtnews.vt.edu/articles/2012/05/051512-caus-aiasolar.html. Accessed 24 Nov 2020

Warmann C (2010), 100 11th Avenue by Jean Nouvel. Deezen.com. https://www.dezeen.com/2010/05/28/100-11th-avenue-by-jean-nouvel-3/. Accessed 26 Nov 2020

Welch A (2020), Al Bahar Towers. Aedas designs world's largest dynamic building façade. Available via e-architect.co.uk https://www.e-architect.com/dubai/al-bahar-towers-abu-dhabi Accessed 21 Dec 2020

Williamson C (2014), May/September: An Interactive Wall Sculpture by Rob Ley, via deisng-milk.com. Available at: https://design-milk.com/mayseptember-interactive-wall-sculpture-rob-ley/ Accessed 15 Jan 2021

Wingårdhs (2011), Kuggen. Available via Archilovers. https://www.archilovers.com/projects/41890/kuggen.html#images Accessed 23 Feb 2021

Wright H (2018), Shanghai, past and present. Available via Abitare.it. https://www.abitare.it/en/architecture/projects/2018/10/21/heatherwick-foster-bund-finance-centre-shanghai/ Accessed 25 Feb 2021

Wurm J (2017), Façade modularisation meets variability: the curtain walling of tomorrow. Available via ARUP.com. https://www.arup.com/projects/futurium Accessed 25 Feb 2021

Wurm J et al. (2020), The Futurium Berlin: Ventilated Façade System with Structurally Bonded Textured Glass. Available via GlassonWeb. https://www.glassonweb.com/article/futurium-berlin-ventilated-façade-system-with-structurally-bonded-textured-glass Accessed 25 Feb 2021

Zacchei V (2011), Il cemento autopulente. Post-it Material Deisgn. http://materialdesign.it/it/post-it/il-cemento-autopulente_13_143.htm. Accessed 24 Nov 2020

Zaha Hadid Architects (2017), King Abdullah Petroleum Studies and Research Centre / Zaha Hadid Architects, via Archdaily.com. Available at: https://www.archdaily.com/882341/king-abdullah-petroleum-studies-and-research-centre-zaha-hadid-architects Accessed 18 Jan 2021

Zaha Hadid Architects, (n.d.) King Abdullah Petroleum Studies and Research Centre. Available at: https://www.zaha-hadid.com/architecture/king-abdullah-petroleum-studies-and-research-centre/ Accessed 18 Jan 2021

# Chapter 5
# Why Windows Matter?

According to the Smart Architecture classification performed in the previous chapter (see Chap. 4), reasons for adopting Smart Advanced Technologies within building envelope are mostly related to the control of solar radiation and outdoor temperature, as figured out by the fact that most of the technical solutions showed in the sample investigated addresses transparent building components.

By assessing external factors associated with smart building technologies' adoption, on the representative sample, it was found that solar radiation and the need of maintaining internal thermal comfort under reference levels are the most frequent determinants since they directly influence the thermal and visual comfort as well as buildings' energy performance. Transparent components have the abilities to almost completely let the solar radiation pass through them, allowing the trigger of physical phenomena induced by incident solar radiation and, as a discontinuity element of building envelope, must fulfill very strict requirements, primarily by the material that composes them (generally the glass), which needs a very implementation in terms of performance.

Despite growth in the application of glass over the years, due to its specific functional role coupled with its aesthetic quality—really appreciated by architects and designers—technical, performance, and maintenance criticalities connected with installation in the building sector still remain.

While a key desirable feature in architecture, windows are recognized as a key weak spot within buildings' energy efficiency.

Additionally, although the building industry is increasingly permeating by advanced and high-performance technologies, the building sector is still absorbing a huge amount of final consumption of the entire country (approximately 40% of the entire energy consumption percentage), generating one-third of the global greenhouse gas emissions (GHG; being responsible for about the 32% of GHG emissions in Europe) (Fig. 5.1).



V. Frighi, *Smart Architecture – A Sustainable Approach for Transparent Building Components Design*, https://doi.org/10.1007/978-3-030-77606-0_5

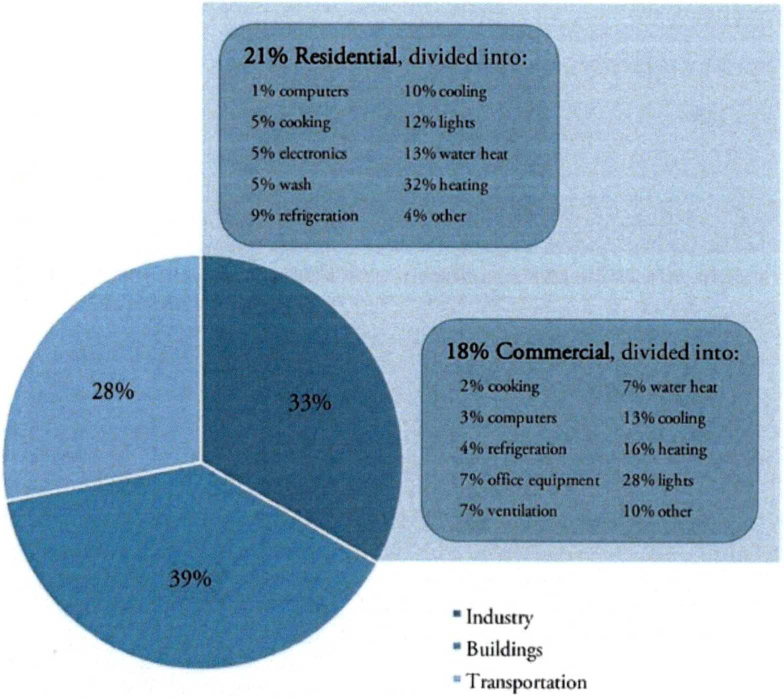


**Fig. 5.1** Fenestration impacts on Building Energy Consumption. Author's elaboration on Building Technologies Program of the U.S. Department of Energy data, Environmental Energy Technologies Division

Although Europe is well on track to meet its targets, doing well with renewables, heat and gain losses through, windows still play a significant role in the energy efficiency of buildings, being responsible for about 60% of the total energy consumption of a building (Jelle et al. 2012) (Fig. 5.2). When compared to insulated walls with the same extension, indeed, the heat gain phenomena on fenestrations have 100 times higher effects (Al-Obaidi et al. 2014). Experimental tests and measurement campaigns recently carried out (Cuce 2017) showed that almost half of these phenomena could be attributed to air leakages and heat transfer mechanism determined by incident solar radiation.

Fenestrations are, by nature, discontinuity points of the building envelope, as openings in opaque building components surrounded by a wooden, metallic, or plastic frame that retains a glazed pane. Glass, when compared to a heavyweight structure like masonry, is a very light material that suffers daily temperature fluctuations; so, its high-transmission properties and, particularly, its solar radiation transmission feature have a great influence on buildings' lighting, heating, and cooling requirements, significantly affecting the amount of energy transmitted through their envelope.

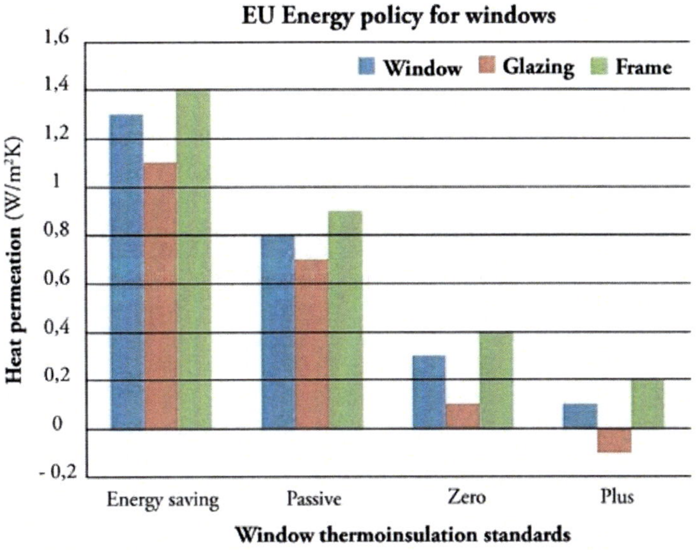


**Fig. 5.2** Incidence of windows' components on the performance of the whole building envelope performance in terms of heat permeation. Author's elaboration on Building Science Consortium data

Even buildings with a small amount of glazed surface, if not adequately protected, would experience overheating in summer and significant heat losses in winter, whose resolution is entirely entrusted to artificial systems. In other words, windows are crucial elements to control the energy performance of the buildings.

Critical issues of windows are *winter heat losses* due to fenestration poor insulating properties since they are made of different and heterogeneous materials and *overheating during summer* due to the amount of entering of solar radiation through the surface of the glass, which can considerably increase the indoor temperature.

In addition to these two main challenges, glazing has the task to limit air infiltrations while allowing natural ventilation, letting daylight to enter for energy cost savings concerning artificial lightning, and permitting occupants to interact with the outside world.

However, despite the growing consciousness about these issues, there is still a lack of an overall strategy able to properly manage the design complexities inherent in transparent building components within building systems, especially if equipped with adaptive features; in addition, there is still a not-so-spread awareness about the fact that they differ from other building components as they can make a positive energy contribution to the whole buildings' performance by letting in free solar energy.

For these reasons, their design optimization challenge is more complex than many other building components and systems. Developing windows with reduced weight and costs and with optimum thermal features, able to control and harvest energy, is of paramount importance as they will have a high impact in the window industry and could be used for both new constructions and for retrofitting of existing buildings. Therefore, proper intervention in the design of glazing systems

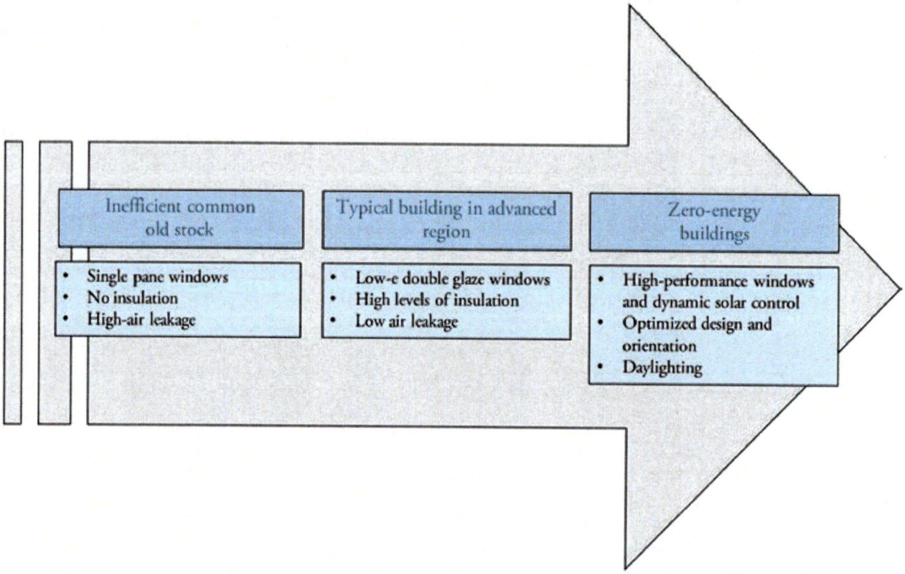


**Fig. 5.3** Building sector needs to change the way in designing, renovating, and building, to shift from an inefficient old building stock toward zero-energy buildings. Author's editing based on Technology Roadmap-Energy efficient building

represents a key opportunity for the building sector to control energy requirements, helping to move toward the goals of the latest energy agenda (Fig. 5.3).

## 5.1 Design Features

The most common material employed to realize transparent building components is *glass* due to its unique characteristic of light transmission.

For thousands of years, artists and architects have worked with it because of how it forms, feels, and handles light; also, craftsmen have used glass for practical applications because of its stability, waterproofing, and transparency (Morse & Evenson 2016).

Even if first glass applications within the building sector appeared around 2000 years ago, used to seal off entrances to structures, the use of glass in buildings did not become widespread until a few centuries ago, with the development of the modern trend in architecture, and it was not until the twentieth century that glass performance began to evolve significantly for residential housing and commercial buildings.

In the last century—since the development of the float glass process in the 1950s—scientists have made extraordinary advances in the characterization and fabrication of glass, leading to innovative applications in diverse fields, such as architecture, transportation, electronics, communications, and medicine (Fig. 5.4).

**Fig. 5.4** Robotized float glass unloading. Source: https://commons.wikimedia.org/wiki/File:Float_Glass_Unloading.jpg by ICAPlants (own work). CC BY-SA 3.0 (https://creativecommons.org/licenses/by-sa/3.0) via Wikimedia Commons

### *5.1.1 General Properties of Glass*

Generally, glazed surfaces suitable for building applications are primarily made of silica ($SiO_2$). However, light nature, as well as transmission/absorption variations determined by different materials, entails the use of different kinds of glasses, with different performance and application characteristics from optical, mechanical, and energetic points of view (Tucci 2006).

Clear common glasses have a light-transmission coefficient included between 60 and 80% for radiations with wavelengths approximately between 400 and 2500 nm, even if intervening on glazing chemical composition, varying its mixture, allow to modify this limit or affecting other performances.

The presence of different chemical components in glass mixture influences the basic performance of glass, such as transparency, light transmission, or selective absorption capacity.

As stated, earlier, the amorphous structure of glass is commonly obtained from the fusion of silica with oxides, inorganic compounds that can be identified as grid- or network former, the so-called *vitrifying*, able to confer chemical stability to final products. The basic product used is silica (silicon dioxide or silicon sand—$SiO_2$ for

69–74%) or boric oxide,[1] generally introduced in the mixture in the form of sand. This compound, through the heat action, gives the glass its texture, transforming it from a crystalline form to a vitreous, amorphous form.

Other compounds are those defined as *grid-* or *network modifiers*, that are:

- *melting agents*, which have the purpose of lowering the melting temperature of silica mass from about 1700 °C to 1550 °C. The more used melting agent is sodium oxide ($Na_2O$ for 10–16%), introduced as sodium carbonate to homogenize the melting mixture of the vitreous paste, even if it can lead to the detriment of glass chemical resistance. Soda can be replaced also with potassium oxide ($K_2O$) or potash, with more energetic melting actions able to make glass more brilliant and colorless, even if facing higher costs. Their use is limited to special glass realizations.
- *stabilizers*, which improve material chemical hardness and resistance. The most common stabilizer is calcium oxide (but aluminum, magnesium, and zinc oxide are also used) and introduced into the mixture as calcium carbonate to limit the solubility of the molten mass, making it less plastic in favor of greater chemical stability and a lower expansion coefficient, thus providing to it better atmospheric agent and mechanical resistance. Also, magnesia (MgO), sulfur dioxide ($SO_2$), and iron ($Fe_2O_3$) can be used as stabilizers.
- *refining agents*, materials that ensure consistency of glass, releasing gases and standardizing its quality, thus adjusting physical and chemical properties of glass avoiding the formation of air bubbles (generally during molting process). They could be sodium sulfate ($NaSO_3$), sodium nitrate ($NaNO_3$), sodium chloride (NaCl), arsenic oxide ($As_2O_3$), calcium fluoride ($CaF_2$), or carbon (C). Compounds that give glasses additional performances are aluminum oxide ($Al_2O_3$), introduced as alumina to extend the workability range as well as chemical and mechanical resistance of material; lead oxide (PbO), to obtain crystal because it confers extreme clarity and high-glare index to glass; and zinc oxide ($ZnO_3$), introduced as a trace to obtain a glass resistance to thermal shocks and chemical corrosion. The introduction of metal compounds dissolved in the mass confers to glass particular selective light absorption properties.
- *matting agents*, which act on transmission properties of the glass.
- *colorants*, able to influence glass' spectral transmission. They can be used also to discolor and optically neutralize the coloration determined by ferrous compounds. The most employed colorants are feldspars (as manganese dioxide—$MnO_2$). Decolorized glasses have a higher energy absorption coefficient with respect to a glass with a high concentration of metal oxides.

Conventional industrial glazing employed in architecture are generally:

- *Soda-lime glass*, produced through the floating process and mainly composed for the 71–75% by silica dioxide ($SiO_2$), for the 12–16% by sodium oxide ($Na_2O$),

[1] The use of boric oxide instead of silica determines the lowering of mass melting point coupled with a greater fluidity of the vitreous paste. Thus determines also a higher resistance toward chemical attacks and a lower expansion coefficient that makes glass suitable to endure high-temperature and thermal shock.

and for the 10–15% lime (CaO) plus a small number of other materials ($Al_2O_3$, CaO, and MgO), employed for adding glass special properties.

- *Quartz glasses*, able to support sudden temperature leaps because of their low expansion coefficient. They are formed almost exclusively of silica and therefore difficult to process. Quartz-glazed products are primarily employed for laboratory objects and applications.
- *Borosilicate glasses*, very resistant to chemical corrosion, have low thermal expansion coefficient and high thermal resistance that increase the shock-resistance ability. Due to their quite high costs, they are rather exceptional in construction.
- *Lead glasses*, very glossy, with high glare index and low softening temperature. To this category belongs crystals.
- *Silica glasses*, which have high thermal and chemical resistance, are characterized by a less silica percentage than quartz glasses. They are employed for special chemical or optical productions because of the difficulty in processing silica.

## Chemical Properties

Chemical resistance of glass generally refers to the performance offered by panes' surfaces, different in each vitreous mass due to the friction with the machinery employed for production processes that can determine small superficial fragments leakages.[2] Compounds that mostly affect chemical resistance of glass are alumina ($Al_2O_3$) and magnesia oxide (MgO).

Chemical properties of glass are also determined by melting time because lower melting times determine inhomogeneous panes at the submicroscopic scale, therefore more susceptible to be borrowed by aggressive chemical compounds.

In general, it can be said that glass chemical resistance depends on the attack to which it is subjected to. When subjected to an acid attack, glass suffers very little damage because the corrosive action does not explicit itself on the siloxane glass chain but through an ion exchange between acid ions H+ and the vitreous ions Na+ and Ca+. Pane's surface becomes poor of alkali superficially present and the corrosive attack can continue if other alkaline ions migrate in the surface. Given that this migration process is very slow, the speed of the whole phenomenon has lowered down over time. The only exception is constituted by the hydrofluoric acid (HF) capable of notching the siloxane chain to cause serious damage to the chemical structure of the glass.

Subjected to an alkaline attack, the OH ions react directly with the siloxane glass' chain, determining the grid breakage up to the formation of silicic and soluble acid anions; this reaction causes the complete dissolution of the glass. Furthermore, its solubility increases the pH of the etching solution. The amount of glass dissolved by an alkaline agent linearly increases with time.

[2] In a pane just laminated, the percentage of sodium present in the surface layers is considerably lower than the one of bulk glass (about 100 Å). In float glass, this difference is even more evident. The two surfaces result both pressed because of alkali- and alkaline-earth oxides loss. On the face in contact with the tin bath there are also missing OH groups, partially replaced through a migration of SnO and $Sn^{2+}$, transferred from the molten metal which causes an imbalance of pane's internal forces because less compressed than the opposite face.

Caused by a sum of the previous attacks, water attack first determines the exchange between H+ ion and the alkaline ion, and, subsequently, due to the increase of the etching solution's pH, the pane undergoes alkaline corrosion. Water attack is dangerous especially in the case of stagnant water on the surface that, as a result of an evaporation process, progressively increases its acidity, attacking glass with an alkaline reaction.

In general, surface corrosion effects occur with iridescent areas whose transparency is characterized by diffusing light effects.

To improve chemical resistance of glass, its surface can be made resistant to corrosive agents by varying the percentages of vitreous mixture compounds, as with the alumina (in calcium–sodium glasses), or by treating it with gaseous $SO_2$ or, again, with fluorinated compounds able to cause the removal of residual alkali from the surface, making the glass less attackable.[3]

Glasses with higher chemical performance are boron-silicate ones, used for producing additive containers for medicines' storage and characterized by chemical neutrality.

## Physical Properties

Glass is a material with excellent heat conduction characteristics.

Heat transmission takes place according to three processes that can be superimposed: conduction, convection, and radiation (Fig. 5.5).

*Conduction* is a phenomenon that occurs between solid substances, in contact with each other, due to a thermal difference. Heat flows from the hot body to the cold body, without mass exchanges but only with energy exchanges at the molecular scale, driven by the thermal excitation. The extent of the phenomenon depends on the material's chemical–physical characteristics.

*Convection* is a phenomenon that occurs between a solid and a fluid (gas), or vice versa, by way of a temperature difference that implies also a movement of the fluid itself. If the movement is determined by the density difference (and therefore by a temperature delta), the convection is defined as natural; otherwise, if the movement is induced by external forces, the convection is defined as forced.

*Radiation* does require physical contact between two bodies; thus, it can also occur in the presence of a vacuum. The energy transmitted by irradiance from a body to another is propagated with electromagnetic waves, with frequency depending on the issuing body temperature. The energy transmitted by a body per unit area and in time unit is called emittance, and it is proportional to the fourth power of the body's absolute temperature.

[3] Surfaces treated in this way are more resistant due to the impossibility to cede additional alkali ions because their superficial layers are devoid of these elements.

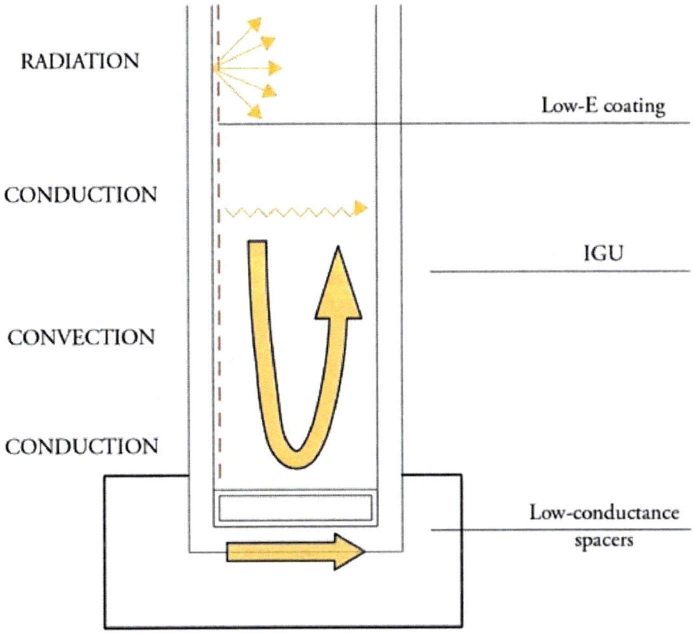


**Fig. 5.5** Heat transmission phenomena that occur through a window system. Author's editing on AGC Glass data

Therefore, the emissive power of a material depends on the material that constitutes it, its temperature, and its geometric-superficial conformation. Rough surfaces, for instance, have generally a greater emissive power than smooth ones. The lower is the emittance, the higher is the reflectance in the long infrared frequency, thus the performance in terms of reduction in thermal losses of building envelope. Materials that have better characteristics of low emissivity are gold, silver, copper, and iridium, whose oxides, if applied as superficially deposit on glazed surfaces, are able to significantly reduce their emissivity.

### Acoustic Properties

The acoustic attenuation offered by glass—even defined sound insulation capacity (SIC) or sound transmission loss—varies according to the frequency considered. SIC is defined as the difference between the sound pressure level in the room where the emission source is placed and the sound pressure level in the disturbed environment.

Glass being a homogeneous material, its SIC is a linear function of the product of the mass of the superficial unit multiplied for the frequency. Due to its high weight, SIC of the glass is quite fair; however, there is performance decay due to the phenomena of resonance, detectable at low frequencies, and of coincidence, that occurs at high frequencies when the sound wavelengths correspond with walls' inflectional wavelengths.

SIC of glass increases when resonance and coincidence phenomena decrease.

In determining the SIC value of a window system, the reduction index of eventual air leaks, the SIC of the frame, the nature of gaskets and sealants, and the overall tightness of the whole fenestration are also involved.

### Mechanical Properties

Glass is a fragile material with different kinds of qualities, for instance, its stress resistance, variable due to loads. Inevitably, its stress resistance decreases while loads and related application time increase.

Generally, the mechanical performance of glass is reduced mainly due to a stress-corrosive phenomenon, linked to environmental humidity that affects glass' silica chain, resulting in the breakage of bonds Si–O–Si.

During the vitrification stage, micro-cracks[4] can occur, responsible for a reduction in glass strength to values between 70 and 350 $N/mm^2$. A further reduction can be determined by major micro-fractures, generally visible, caused by mechanical shocks during processing phases.

Mechanical resistance of the glass depends also on the superficial quality as well as the size of the stressed area of the glass. This feature can be corrected or improved with tempering or stratification processes that involve more consisting layers of the pane, with cleaning processes that treat the outer surface of the glass with hydrofluoric acid solutions capable to significantly reinforce it.

In float glasses, the imbalance of mechanical performances between the two panes determines the different degrees of compression, which means different mechanical resistance between the two surfaces: The less compressed surface will have greater mechanical resistance, while the more compressed determines a further lowering of the admissible breaking load.

Additional mechanical requirements are those connected with wind- or snow-stress resistance and safety or features such as thermal shock resistance. Even temperature differences in distinct points of the plate, determined by nonuniform irradiation, can determine abnormal strains, sometimes higher than those admissible (Table 5.1).

Thanks to the advancement in technological progress and industrial techniques, glass is nowadays available in various forms and compositions, generally distinguished based on the manufacturing process that generated them.

[4]Those micro-cracks (of elliptical form with microscopic axial length) are capable of providing in a fragile body such as the glass' pane the trigger of fractures because tensile stresses that are concentrated thus exceed acceptable ones, bringing the pane toward its breakage.

**Table 5.1** Mechanical and thermal features of the Main glass. Source: UNI EN 572-1 Glass in building—Basic soda lime silicate glass products—Part 1: Definitions and general physical and mechanical properties

| Properties | Symbol | Unit of measurement | Value |
|---|---|---|---|
| Density (at 18 °C) | $\rho$ | [kg/m$^3$] | 2500 |
| Hardness (Knoop) | $HJ_{0.1/20}$ | Gpa | 6 |
| Young module (elasticity module) | E | [MPa] | 70,000 |
| Poisson coefficient | $\mu$ | [−] | 0.2 |
| Bend resistance | $f_{g,k}$ | [MPa] | 45 |
| Specific thermal capacity | C | [J/(kgK)] | 720 |
| Average linear expansion coefficient (20–300 °C) | $\alpha$ | [1/K] | $9 \times 10^{-6}$ |
| Resistance towards the temperature gap | | [K] | 40 |
| Thermal conductivity | $\lambda$ | [W/(mK)] | 1 |
| Average refraction index towards visible radiation | N | [−] | 1.5 |
| Emissivity | $\varepsilon$ | [−] | 0.837 |

### 5.1.2 Window Panes

Despite the classification of glass employable in fenestration systems from an energetic point of view, window panes can be divided according to their manufacturing process, distinguished as *basic products*, intended as those subjected only to one manufacturing (such as annealed glass), and *secondary manufactured products*, such as toughened glass, laminated glass,[5] and coated glass.

Other glazing products are listed subsequently, according to the manufacturing they are subjected to.

*Pressed Glass* products are obtained by pressing and hot molding the liquefied glass paste into a metal mold. Through this manufacturing process, glass bricks can be obtained; they are building products able to guarantee structural performance, a distinctive feature of opaque walls, without sacrificing light transmission of the glass. Pressed glass' products can be formed as a massive slab, hand-woven with multiprismatic surfaces, hollows, and air chambers. With this technology colored paste vitreous tiles for floors and claddings can be obtained to be put in place with adhesives or special mortars.

*Blown Glass* is produced with ancient traditional techniques (mouth-blown) that allow obtaining products with a wide range of shades and chromatic intensity. Blown glass products are often characterized by imperfections due to the craftsmanship of the processing; the few plates produced with this technology are mainly used in art glasses.

[5] Made of two or more panes of glass sealed together with an interposed plastic layer (typically a PVB, which has qualities of transparency, elasticity, toughness, and cohesion to the glass panes) that can prevent the fallout of glass shards after fractures. The presence of the plastic layer as well as the alternation of different components' thicknesses provides also protection from UV rays and gives the system good acoustic feature.

*Sheet Glass* is obtained by the mechanical circulation of the molten mass. This kind of glass presents discontinuous thickness due to undulations of their surfaces, more or less accentuated because of the processing technique that can determine optical deformation of glass. Today, this processing is out of practice, and it is replaced by the float technology that allows producing glass with better physical–chemical properties and identical light transmission factors.

*Molten Glass* (also defined *laminated glass*) is produced by the continuous drain of the vitreous mass from a basin oven by rolling on a thin layer of sand. This technology is used for producing slabs of great thickness. Molten glass products are opal and translucent crystals.

*Wired Glass* constitutes a subfamily of molten glass. Wired glass—produced by draining—transmits light in a diffused way. It incorporates, within the mass of a fluid glass, a welded steel mesh (with real armor function) that helps retaining glassy fragments consequent to a possible breakage of the plate. This kind of product has high fire performance while its downsides are low mechanical strength—solvable intervening on glass–metal bond—and the possible presence of gas bubbles within the slab.[6]

*Profiled U-Glass* (or *Channeled Glass*). Commercially called U-GLAS, profiled or channeled glass is produced as drained glass: The glass ribbon, after the melting tank, is conducted on molds that bend upward its edges. The final product has a resistant self-supporting U-section that permits its fastening directly on supporting structures, without needing any other metal supporting frames, thus allowing to obtain large glass surfaces. Profiled glass can also be reinforced with stainless steel wires—placed longitudinally—to increase the mechanical strength of the plate. The seal between planks is ensured by the use of plastic or silicone structural sealants.

*Screen-printed Glass* is panes decorated with drawings, signs, or graphics of various types obtained through photographic reproduction techniques using the construction of a screen-printing frame. The engraved image is then fixed through a vitrification process induced by high-temperature firing.

*Ground Glass* (or *Etched Glass*) is obtained through a process that confers translucency properties to the glass pane using two different techniques: sanding and polishing. In the first case, the glazed surface is machined with a sand jet that makes it opaque and rough; in the second case, the pane is treated with acids that confer variable degrees of opacity (Fig. 5.6).

Talking in performance terms, glass can be used for fenestration purposes as a *single pane* or in a *multilayer solution*, otherwise called Insulated Glass Unit (IGU). While single-pane solutions are, by now, obsolete—because they are not able to guarantee minimum regulations' standards—IGUs are even more widespread, thanks to their cost-effectiveness (Roos and Karlsson 1994).

---

[6] To fix this problem can be employed low carbon steels due to the contents of pearlite and cementite in their structure able to avoid the formation of gas bubbles.

| Raw glass | Processed glass |
|---|---|
| • Float glass | • Insulating glass |
| • Laminated safety glass | • Heat-treated glass |
| • Coated glass | • Chemically toughened glass |
| • Mirror-Painted glass | • Laminated safety glass |
| • Patterned glass & wired patterned glass | • Enamelled glass |
| • Acid-etched glass | • Silk-screen printed glass |
| | • Active glass (BIPV) |
| | • Surface treatment |
| | • Fire-resistant glass |

**Fig. 5.6** Overview of the main raw glass and processed glass products. Author's editing on AGC Glass data

IGUs are basically glazed panels made by two or more panes sealed together along the perimeter and filled in the interspace between them with fixed air or low-emissive gas. Different kinds of IGUs are identifiable depending on (a) the type of glass panes employed; (b) the frame materials; (c) the width of the interspace; (d) the nature of the gas contained in it; (e) the desiccant inserted in the spacer; and (f) the type of sealant employed (Rezaei et al. 2017) (Fig. 5.7).

The thermal performance of such systems is determined by the number and type of panes and the gas inserted between them; interspace plays a significant role in reducing heat exchanges through the glazed surfaces, even if there is a maximum dimension beyond which performance improvement is not remarkable. As a matter of fact, over a certain dimension, gas convective motions within the interspace would induce heat losses; this is the reason why the employed insulating gases are generally heavier than air and with a lower tendency to move (e.g., Argon), capable of slowing down heat flow through the glass pane from inside to outside (Karabay and Arici 2012).

Moreover, the duration and, consequently, the effectiveness of an IGU are strictly connected with sealants' quality, often a weak point of the system capable of affecting the insulation provided by the air gap. Two types of sealants are generally used: *two-component sealants*, which chemically react as a result of their application, and *sealants applicable in a stable form*.

In general, sealants have to be elastic enough to allow mutual movements between plates and the spacer, subjected to thermal variations or mechanical stresses as well as their chemical components must be compatible with gases contained in the interspace. To neutralize vapors that may persist inside the interspace during the assembly phase, or which may seep later through sealants—and thus capable to affect the glazing characteristics with condensation phenomena—dehydrated salts are inserted in the spacer with the task of selectively absorbing the gas contained in the gap. Generally, salts used are silica gels or molecular metals.

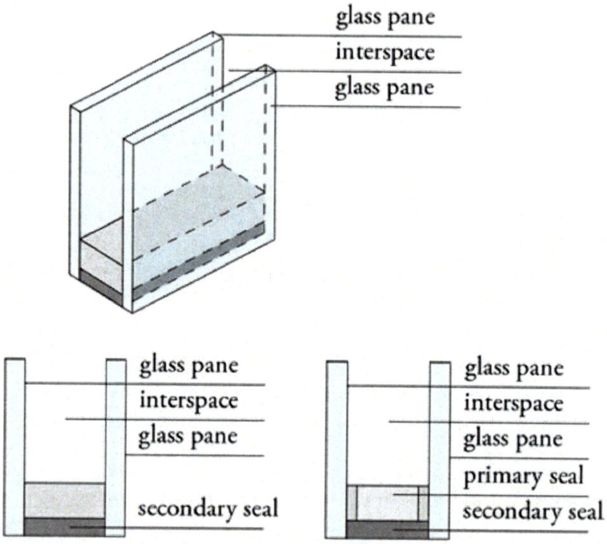


**Fig. 5.7** Composition and geometry of a double-pane IG unit. For further insights, refer to Van Den Bergh et al. (2013)

**Fig. 5.8** Different kinds of windows frames: from left to right, a wooden frame, a PVC frame, and a metal frame

## 5.1.3 Frames, Spacers, and Other Devices

Significantly responsible for the thermal behavior of the whole window performance, frames constitute up to 30% of the exposed surface of fenestration (Conato and Frighi 2016).

Existing frames can be made of various materials, both in single and in mixed solutions. Major of fenestration frames are today made of wood (55%), followed by aluminum (24%), PVC (11), and other material solutions (Fig. 5.8).

The traditional frame material is wood due to its good insulation properties[7] coupled with a fair mechanical resistance, even if with some maintenance issues (in the

[7] Frame U-factor stands in the range of 1.0–2.80 W/m$^2$ K.

untreated version). A compromise for obtaining a frame with the same aesthetic quality as wood one could be a two-component frame, in which wood clads the interior face while aluminum or vinyl faces the outside, creating a permanent weather-resistant surface.

However, these mixed solutions need expedients to avoid thermal losses through them due to the combination of two different materials. These resources, called *thermal break solutions*, consist of the insertion of plastic material with low thermal conductivity within the frame section, to divide the cold side from the hot side, thus avoiding the thermal transmission between indoor spaces and the outside.[8] Sometimes, thermal breaks present cavities that, in turn, can be filled with insulation materials, increasing in this way their action (Fig. 5.9).

The best window frames normally employ materials with a low thermal conductivity as thermal breaks, reaching U-factors between 0.6 and 0.8 W/m$^2$ K. Further steps foresee the integration of innovative materials (such as aerogel) within frames to improve thermal breaks features. Inserting aerogel bars inside an aluminum frame allows achieving a U-value of 0.57 W/m$^2$ K (Casini 2016).

Another variation could be obtained with a composite frame, in which wood particle and resins are compressed and extruded into shapes for window frame; the advantage of such technology is great stability, coupled with structural and thermal properties even better than those of a conventional wooden frame, with better moisture resistance and more decay resistance; in addition, they can be textured or painted as desired.

Other employed materials are metals used for their resistance, lightness, and durability (especially if anodized), coupled with the ease to extrude them in shapes required for windows' parts.

These solutions are certainly less qualitative in terms of thermal insulation when compared to wooden frames unless they provide thermal break. Generally, aluminum frames are utilized in large elements due to their good stability, cost-effectiveness, and low maintenance needs. The different superficial treatments can compensate for the irregular oxidation of the natural material.

Steel frames, on the other hand, are commonly used in industrial applications even if they present few installation problems due to their extremely high weight.

Other common solutions are plastic frames made of polyvinyl chloride (PVC), a light, convenient, and economical plastic material with good insulating value. PVC frames have a hollow section (with single or multichamber) with various shapes that help in reducing convection exchanges; they can include or not a metal stiffener to strengthen their mechanical resistance. Negative issues of such technologies, on the other hand, can be their dimensional stability and the durability of their whiteness over time.

---

[8] Current technology with standard thermal breaks has decreased aluminum frame U-factors from roughly 11.35 to about 5.6 W/m$^2$ K.

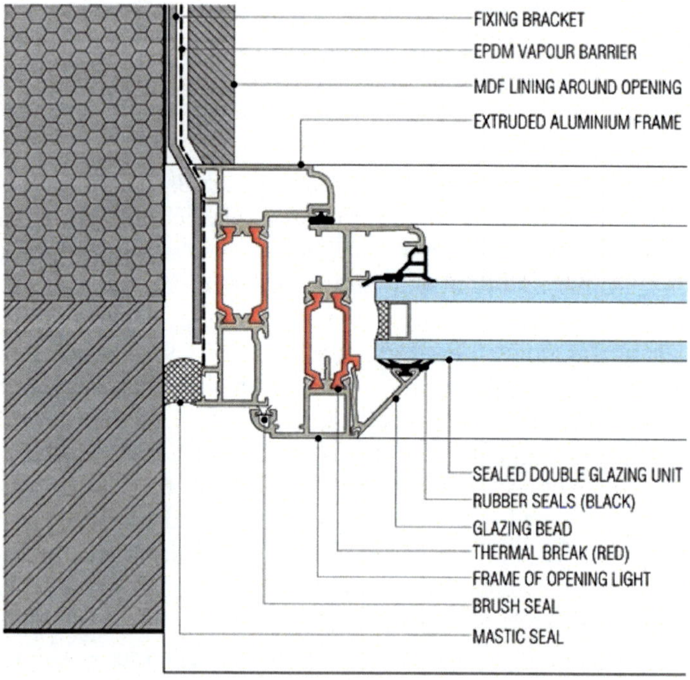


**Fig. 5.9** Vertical section of an extruded aluminum frame with thermal break. Source: https://commons.wikimedia.org/wiki/File:Aluminium_window_detail.png by ProfDEH. CC BY-SA 3.0 (https://creativecommons.org/licenses/by-sa/3.0) via Wikimedia Commons

Regarding spacers, two kinds of spacers exist: *traditional solutions*, made of wood, aluminum, and steel, or, on the other hand, *solutions with improved thermal performances*.

A spacer is a device used to join the glass panes along the perimeter whose main function is to determine the width of the interspace between the two glass sheets, avoiding the formation of condensation using, within it, a desiccant agent (generally dehydrated salts).

In addition, the spacer system accommodates stress induced by thermal expansion and pressure differences to provide a moisture barrier that prevents the passage of water (or water vapor that would fog the unit) as well as to provide a gas-tight seal that prevents the loss of any special low-conductance gas in the air space.

Standard spacer solutions have a hollow section and are generally made of metal (typically aluminum), even if stainless steel spacers exist and have lower thermal conductivity and, sometimes, different cross-sectional shapes.

Another approach is to replace metal with better insulating materials, such as foam or thermoplastic resins. Foam spacers (pre-desiccated and made of structural foam) are generally equipped with improved thermal properties, compared to aluminum ones, and are defined *Warm Edge Spacers* (Elmahdy 2003). The more significant benefit resulting from its application is the increase in interior surface temperature (at the window's bottom edge) which is most subject to condensation.

Finally, many other products and components useful for assuring correct laying operation and installing of window systems are available on the market in multiple ways and with several features according to different application contexts.

Nevertheless, the general common devices that can be recognized in almost all current installations are:

- the *counterframe*, employed to fasten through its application the window frame to the wall square
- the *windows' ledge*, used where necessary to appropriately finish the bottom edge of the interface and to avoid water penetration that can compromise installation
- all the devices employed to ensure occupants' privacy, such as *blinds*, *roller shutter*, and *solar control devices* in general.

To ensure a correct installation, components that try to integrate with a single product multiple functionalities have recently been developed to avoid the criticalities that the interface between heterogeneous components could imply.

Generally, the element that gathers the whole abovementioned issues is the counterframe, a connection element between the window and the wall. Counterframes spread in current building practice are of different typologies because they have to adapt to different needs and shapes as well as to the various types of windows adopted. The most traditional solution is the one that employs counterframes entirely made of wood simply joint together with metallic nails; they are generally fastened to the wall through metal brackets and fixed with dowels, even if different kinds of fastening can be adopted according to the type of wall support. Other common counterframe types use metal, PVC, or composite materials; recently, to overcome potential critical issues related to the adoption of such basic solutions, new improved versions of them have been developed. Generally defined as *mono-block*, these prefabricated systems—already thermo-insulated—allow resolving the whole interface, simplifying the installation phase of a window, thanks to the several advantages they offer, such as simplified design phases, enhanced reliability during installation, compatibility with other devices, effective insulation, and elimination of geometrical thermal bridges due to the continuity of the insulation layer on the external facade of the wall (Fig. 5.10).

Generally, mono-blocks present a jamb shape to embrace gaskets and wires to avoid crevices. In addition, they help to align the stile to the block, contributing in a significant way to the regulation of the wall square.

Clearly, the main advantage of such a system is its ability to integrate into a single component different functions, such as shading systems, screens, and other devices, resolving, at the same time, the dimensional issue caused by the presence of two distinct elements and ensuring a valuable thermal and acoustic insulation level.

With the same principles, ledges and stiles of windows are existing, generally made of insulating expanded polystyrene (high-density EPS) to avoid thermal bridges in correspondence of such interface, allowing a durable and resistant connection coupled with excellent insulation properties and simple and fast installation.

**Fig. 5.10** Preassembled mono-blocks for windows during the construction phase. Photo credit: the author

To obtain a proper window installation, sealants, gaskets, fastenings, and accessory products are essential.

As a general rule, components and window accessory products must be chosen according to the field of application as well as to their intended use, the environmental stresses to which they will be exposed, and the type of users that will manage them, and their selection must be consistent with the design of the building envelope system in its whole.

## 5.2 Energy Performance Features

### *5.2.1 Introduction to Light and Color*

The first and main feature of glass is its light transmission, allowed by transparency and determined by the interactions between light photons and the atomic glass' structure, able to modify it between reflection and absorption phenomena.

In general, glazed support transmits most of the incident solar radiation, depending on the substances that compose it and/or the surface treatments it is subject to.

The sun, the main power source of radiant energy, generates light and heat in the form of electromagnetic waves, characterized by a wavelength (λ) and a wave amplitude (Δ).

Light, coupled with thermal radiations, represents more than 90% of the energy transmitted by the entire solar spectrum, producing, in turn, a set of radiation of wavelength between 0.3 and 3 μ. Within this amount, only a narrow band of solar rays reach the earth's surface, that is between 380 nm—*ultraviolet*—and 780 nm—*infrared*—visible to the naked eye, which does not perceive radiations with lower wavelength (Table 5.2; Fig. 5.11).

Within the whole solar spectrum, three sections are those that affect internal building environments, being transmitted by the glass; they are Ultra Violet (UV), Visible Light (VL), and Infrared Energy (IE), that lie between the visible-section and the microwave-section of the electromagnetic spectrum. UV light is, in turn, separable into three categories; two of them are rejected by the earth's atmosphere and float glass.

UV-a is the only amount of energy able to pass through it, primarily responsible for fading and deterioration of interior furniture, even if it can be blocked through coatings or films able to reflect or absorb this spectrum.

Infrared energy (IE) represents the radiant heat energy emitted by the sun added to an internal environment; to act on this transmission means reducing the heat inside rooms, thus preventing possible overheating during the summer season. The more IE a glass pane is able to prevent, the more energy and cost can be saved.

Finally, Visible Light (VL) is the amount of light visible to the human eye and consists of natural daylight; it can generate problems connected to unwanted glare and eye strain. Generally, windows can reduce VL by acting on tints.

With respect to the visible wavelengths, glass is almost completely transparent while in respect to radiations with wavelength superior to 3 μ (infrared), it is opaque; this is the cause of the so-called *greenhouse effect* due to which bodies located in a space protected by glass surfaces increase their temperature as a result of direct irradiation, re-emitting energy converted into sensible heat as infrared radiations that remain confined within the space.

In general, incident infrared solar radiation on the surface of the glass can be *reflected*, *transmitted*, or *absorbed*.

In normal incidence conditions, the reflected energy is the quantity of solar radiation reflected back by the glass into the atmosphere; the transmitted energy is the amount of solar radiation directly transmitted through surface of the glass, while the absorbed amount expresses the quantity of solar radiation absorbed by glass, responsible for its temperature's increase. Furthermore, the pane will redraw the energy stored, partially inward and partially outward depending on boundary conditions.

**Table 5.2** Solar spectrum composition

| Type of radiation | Wavelength (nm) | Proportion of energy |
|---|---|---|
| UV | 280–380 | Approx. 5% |
| Visible | 380–780 | Approx. 50% |
| IR | 780–2500 | Approx. 45% |

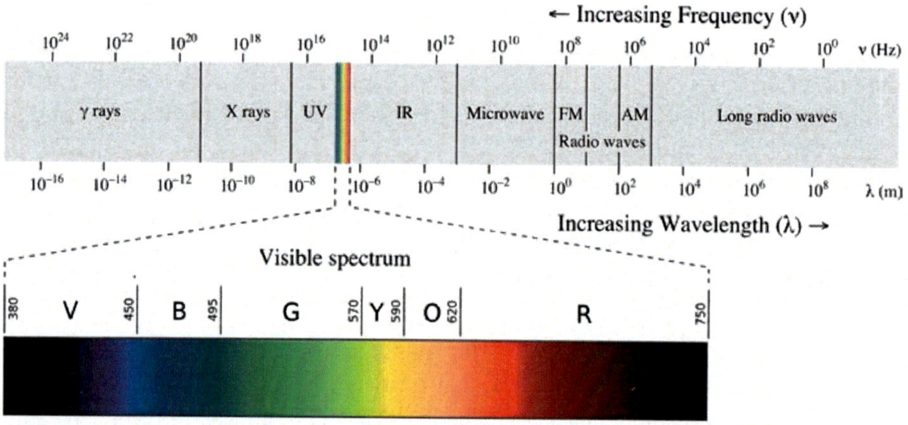


**Fig. 5.11** Distribution of electromagnetic waves with respect to frequency and wavelength, highlighting visible part of the electromagnetic spectrum. Source: https://commons.wikimedia.org/wiki/File:EM_spectrumrevised.png by Philip Ronan, Gringer. CC BY-SA 3.0 (https://creativecommons.org/licenses/by-sa/3.0) via Wikimedia Commons

The relations between the abovementioned energy quantities are expressed through the *reflection coefficient* (r or $\rho$), the *transmission coefficient* (t or $\tau$), and the *absorption coefficient* ($\alpha$), whose sum determines the total amount of the incident solar energy.

The reflection coefficient expresses the ratio between the reflected energy flow and the incident energy flow; in the same way, absorption and transmission coefficients express the relationship between the absorbed and transmitted flow and the incident flow. The sum of these three factors is equal to one.

Total transmission represents the total amount of solar radiation that, in normal incidence conditions, is transmitted through the glass; it is composed of the direct transmission (short-wave component) and of the component dissipated inward for radiation and convection (long-wave component).

The ability of a surface to absorb or emit electromagnetic radiation is reflected by the *emissivity value*.

Due to its nature, glass has a high emissivity. Particular coatings are existing, able to act on this feature, reflecting heat back inside buildings and thus reducing heat losses through windows. With a Low-E coating on its surface, glass does not absorb energetic radiation through its reflection inside the room, reducing heat transfer through the glass and increasing in this way the U-value of the system.

### 5.2.2 *Reference Parameters for Performance Evaluation of Windows*

So far, the transparency feature of glass is a function of the interaction of light photons with the material's atomic structure. This property is the one that allows visible light transmission as a consequence of glass direct irradiations. Besides, radiant

| Heat-transmission through Transparent Building Components | | |
|---|---|---|
| Heat-transfer | from the inside surfaces towards glass surface | Transmission mode that occurs whenever the inner glass' surface has a lower temperature than the internal operating temperature.<br>Heat is transferred:<br>• by radiation between the internal walls' surfaces and the glass' surface;<br>• by conduction and convection of air in contact with inner glass' surface. |
| | through glass | Transmission that occurs when the single glass pane is not able to offer adequate resistance to heat-transfer. By the addition of an interspace – filled with air – it is possible to reduce such heat-exchange due to the air low-conductivity compared to glass' one.<br>The resistance can also be increased by the increasing of the interspace up to a limit beyond which occur convective motions able to make self-defeating thickness increase.<br>The use of low-e coatings allows to reduce radiative exchanges though glass; interposing them in fact it is possible to reduce such exchange of about three quarters of its entity. |
| | from the outer glass' surface to the external environment | Transmission process that occurs by radiation, conduction and convection. It depends on the external weather conditions, on wind-exposure and on the amount of rain precipitation to which it is subjected. |

**Fig. 5.12** Heat transmission modes through transparent building components

energy emitted by the sun as electromagnetic waves comprises heat transmission in the form of infrared energy.

Generally, glass transmits most of the incident solar radiation depending on its composition and on the surface treatments to which it is subjected. While absorption or reflection abilities depend on features of glass, transmission through transparent building components is highly influenced by boundary conditions and occurs by means of conduction, convection, and radiation phenomena (Figs. 5.12 and 5.13).

Therefore, the definition and description of the energy performance of glazed components are quite complex because they can interact with incident solar radiation. For these reasons, transparent building components must be endowed with precise requirements and, accordingly with Cuce and Riffat (2015), have to be described in terms of both thermal and optical parameters.

Basically, the parameters through which a glazed component is evaluated, in performance terms, attained to *thermal control*—expressed through the thermal insulation capacity of the system—and to the *lighting* and *solar control*, expressed through the ability of the glass in contrasting lighting and heating flux, thus controlling the amount of solar energy entering in rooms.

The relative reference parameters commonly used in window rating methods influence the total energy consumption and the indoor environmental comfort:

- *Thermal transmittance*, or Heat Transfer Coefficient (also called U-factor), is the rate at which a window conducts nonsolar heat flow. The lower the U-factor, the more energy-efficient the window is. Currently, over 40% of fenestration systems in the EU have thermal transmittance higher than 5.5 $W/m^2 K$,[9] thus resulting in important heat losses through glazing.

[9] TNO Built Environment and Geosciences, "Glazing type distribution in the EU building stock," TNO Report TNO-60-DTM-201-0038. The Hague (NL):TNO; 2011

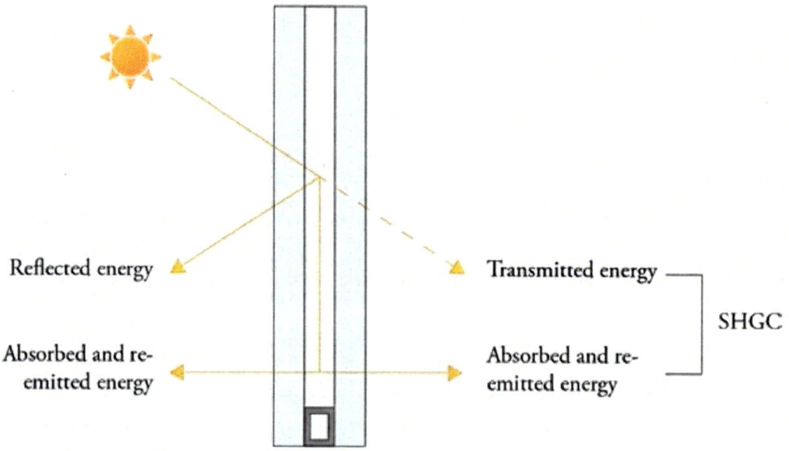


**Fig. 5.13** Light transmitted, reflected, and absorbed through a common glass pane

**Table 5.3** Minimum thermal transmittance for fenestration in case of energy refurbishment, as stated in Appendix B, Table 4 of the Italian Decrees 26th of June 2015 "Buildings minimum requirements"

| Climatic zone | Fenestration thermal transmittance U ($W/m^2$ K) (from July 1st, 2015) | Fenestration thermal transmittance U ($W/m^2$ K) (from January 1st, 2021) |
|---|---|---|
| A e B | 3.2 | 3 |
| C | 2.4 | 2.0 |
| D | 2.1 | 1.8 |
| E | 1.9 | 1.4 |
| F | 1.7 | 1.0 |

- *Solar Heat Gain Coefficient* (SHGC), or g-value. It expresses the fraction of incident solar radiation admitted through a fenestration product, either directly transmitted or absorbed and subsequently released as heat inside an indoor environment. The lower the SHGC, the less solar heat the glazed component transmits, and the greater is its shading ability. Fenestration products with high SHGC are more effective in collecting solar heat during winter, while windows with low SHGC are more effective in reducing cooling loads during summer by blocking heat gain from the sun.
- *Visible Light Transmittance*, or light-transmission coefficient ($\tau$l), is the fraction of the visible sunlight spectrum perceived by the human eye and transmitted through the glazing. It is expressed by a number comprised between 0 and 1: a high VLT value means more visible light transmitted (Tables 5.3 and 5.4).

Another important parameter to evaluate the energy performance of glazed systems is the *Light to Solar Gain* ratio, which measures the spectral selectivity of a glass system as an expression of the efficiency of different glass types in transmitting daylight while blocking heat gains. It is expressed by the ratio of the VLT to

**Table 5.4** Maximum SHGC allowed for building under energy refurbishment, as stated in Appendix B, Table 4 of the Italian Decrees June 26th, 2015 "Buildings minimum requirements"

| Climatic zone | D.P.R. 59/2009 | Decree 26 of June 2015 |
|---|---|---|
| All zones | g (SHGC) = 0.50 | $g_{gl+sh} = 0.35$ |

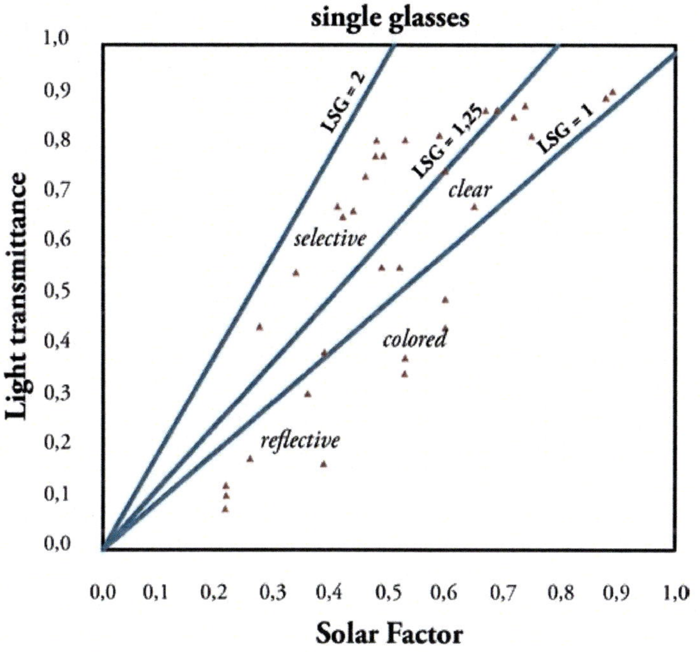


**Fig. 5.14** LSG value for different glazed technologies. For further insights see Van Den Bergh et al. (2013)

SHGC; the higher the number, the more light is transmitted without adding excessive amounts of heat. A higher LSG ratio indeed means sunlight entering the room is more efficient for daylighting, especially in summer where more light is desired with less solar gain. So, the higher the LSG number, the brighter the room is without adding an excessive amount of heat (Fig. 5.14).

## References

Al-Obaidi KM et al. (2014), A Review of Skylight Glazing Materials in Architectural Designs for a Better Indoor Environment. Modern Applied Science 8(1)

Casini M (2016), Smart Buildings. Advanced Materials and nanotechnology to Improve Energy-Efficiency and Environmental Performance. Woodhead Publishing, Cambridge

Conato F, Frighi V (2016), Metodi della Progettazione Ambientale. Approccio integrato multi-scala per il controllo della qualità ambientale. Franco Angeli Editore, Milano

Cuce E (2017), Role of airtightness in energy loss from windows: Experimental results from in-situ tests. Energy and Buildings 139:449-455. https://doi.org/10.1016/j.enbuild.2017.01.027

Cuce E, Riffat SB (2015), A state-of-the-art review on innovative glazing technologies. Renewable and Sustainable Energy Reviews 41: 695–714. https://doi.org/10.1016/j.rser.2014.08.084

Elmahdy AH (2003), Effects of improved spacer bar design on window performance. Construction Technology Update, 58, National Research Council of Canada. https://nrc-publications.canada.ca/fra/voir/td/?id=6f0183ea-6905-4071-9893-b81a2b4a2ca2 Accessed 3 Feb 2021

Jelle BP et al. (2012), Fenestration of Today and Tomorrow: A State-of-the-Art Review and Future Opportunities. Solar Energy Materials & Solar Cells 96:1-28

Karabay H, Arici M (2012), Multiple pane window applications in various climatic regions of Turkey. Energy Build. 45:67-71. https://doi.org/10.1016/j.enbuild.2011.10.020

Morse DL, Evenson JW (2016), Welcome to the Glass Age. International Journal of Applied Glass Science, 7 (4): 409-412

Rezaei SD, Shannigrahi S, Ramakrishna S (2017), A review of conventional, advanced, and smart glazing technologies and materials for improving indoor environment. Solar Energy Materials & Solar Cells 159:26-51. https://doi.org/10.1016/j.solmat.2016.08.026

Roos A, Karlsson B (1994), Optical and thermal characterization of multiple glazed windows with low U-values. Sol. Energy 52:315-325. https://doi.org/10.1016/0038-092X(94)90138-4

TNO Built Environment and Geosciences (2011), Glazing type distribution in the EU building stock, TNO Report TNO-60-DTM-201-0038. TNO, The Hague (NL)

Tucci F (2006), Involucro Ben Temperato. Efficienza energetica ed ecologica in architet-tura attraverso la pelle degli edifice. Alinea Editrice, Firenze

Van Den Bergh S et al. (2013), Window Spacers and Edge Seals in Insulating Glass Units: A State-Of-The-Art- Review and Future Perspectives. Energy and Buildings 58:263-280

# Chapter 6
# Glazing for Smart Architecture

The technological progress that led to the enhancement of glass production techniques, coupled with the development of applied research on innovative materials, has allowed the realization of new products with extremely high performance—when compared to the basic glass product—to overcome the simplification of traditional elements.

However, even if buildings with a great amount of glazed surface are often been considered "green" or "sustainable," these constructions are actually very energivorous, encountering severe indoor climate problems and large energy bills.

It is an undeniable fact that glazing still constitutes a clear design problem, which requires the balance of thermal comfort, energy efficiency, light quality, view, daylight, and relationship with the outdoor environment.

To face the abovementioned well-known challenges, glazed systems with different abilities have been developed, transforming glass, a material that once performed only the function to allow the entrance of light, into a dynamic filter that plays a more active role in regulating internal building conditions.

Special glasses have been made possible, thanks to the evolution of different production techniques that allows to effectively act on properties such as solar radiation control, thermal insulation, passive solar heat gain, glare control, privacy, maximization of natural lighting, and other parameters responsible for internal comfort and energy costs and consumption.

Several types of experimental, innovative glasses that cover a wide range of functions and application cases are today available on the market or, at least, are under development. This group includes the so-called *dynamic glazing,*[1]

[1] A definition of dynamic glazing products comes from the *Energy Star product specification for Residential Windows, Doors and Skylights. Eligibility Criteria*, Version 6.0, that recognize as dynamic «any fenestration product that has the fully reversible ability to change its performance properties, including U-Factor, Solar Heat Gain Coefficient (SHGC), or Visual Transmittances.



V. Frighi, *Smart Architecture – A Sustainable Approach for Transparent Building Components Design*, https://doi.org/10.1007/978-3-030-77606-0_6

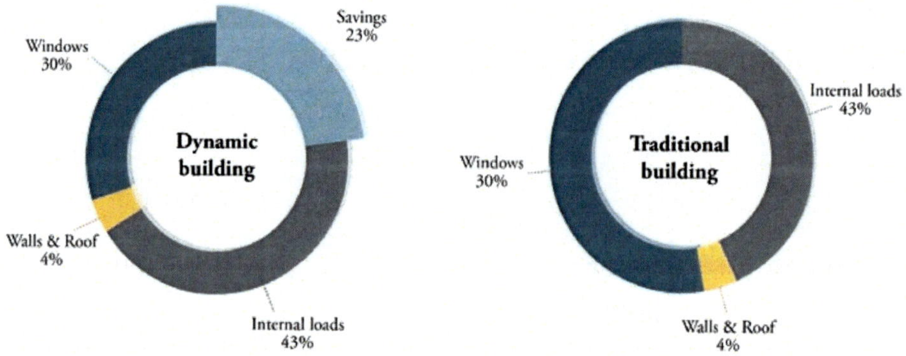


**Fig. 6.1** Peak load comparison between a traditional building with Low-E glazing installed, and a building in which fenestrations are dynamic glasses: in dynamic building, cooling peak loads are reduced by 23%. Courtesy of View, Inc.

fenestration products able to change their performance properties (Allen et al. 2017) by modifying their characteristics of transparency, gloss, coloration, and solar radiation screen while keeping unchanged glass structural properties.

Dynamic glasses, otherwise defined as *Smart Windows,*[2] can be controlled through a variety of means, such as automatic photosensors and motion detectors, smartphone applications, or integration with intelligent building systems, allowing end-users to control daylighting inside rooms by switching their aspect, managing the control of the interactions that take place between the external conditions and the internal environment, and creating, in this way, smart adaptive building envelope with the ability to save costs for heating, cooling, and lighting (Fig. 6.1).

Therefore, it is clear that the advantages deriving from the massive adoption in the building sector of this kind of technologies could lead to significant reductions in energy consumption and through the elimination of all the attached devices needed to control incident solar radiation, also in costs, maximizing at its best natural daylight while minimizing glare and heating–cooling requirements.

To make a quick but rather comprehensive overview about these innovative devices, they have been described and classified according to their performance behavior, thus dividing them into *passive*, if equipped with mainly static performance, or *active*, if dynamically responds to the external input of various nature, glazing technologies. Also, emerging technological development has been taken into account.

---

This includes, but is not limited to, shading systems between the glazing layers and Chromogenic Glazing».

[2]Wikipedia defines smart or switchable those windows whose light transmission properties are altered when when voltage, light or heat is applied. Generally, glass changes from translucent to transparent, changing from blocking some wavelengths of light to letting light pass through.

## 6.1 Passive | Static Glazing Technologies

### *6.1.1 Tinted Glass*

Tinted glass is a mass-colored glass obtained through the addition of metal oxides to a base-clear product during the production process.

The applied oxides can vary the color and optical properties of glass, conferring to the pane the property to absorb certain portions of incident solar radiation, thereby reducing the percentage that passes through it (Fig. 6.2).

Usually, the coloration is obtained with ionic or phase-dispersed colorants made of particle aggregates that determine the attenuation of transparency, caused by diffraction and reflection phenomena of light due to the particles dispersed in the vitreous mass.

An absorbent tinted glass with discrete properties should intercept the energy of the solar spectrum between 800 and 2000 nm, as a result of glass chromatic alteration, without excessively lowering visible transmission (Table 6.1).

The adoption of tinted glass is quite effective in reducing the amount of incident solar radiation entering within a room: According to the results obtained by Chow et al. (2010), thermal transmission in a room with tinted glazing can be reduced by more than 20%; their effectiveness and color-intensity grow proportionally with pane's thickness and, in an inversely way, with its light transmission.

Due to its selective capacity, the high absorption coefficient of these products can easily cause heating of the slab, thus weakening its mechanical strength.

Because of their modest energy selective capacity, absorbent glasses can be coupled with reflective metal coatings to allow the combination between absorbent and selective abilities.

**Fig. 6.2** Pilkington Arctic Blue application. Courtesy of Pilkington Poland©

**Table 6.1** Specifications of most common tinted glass. Author's editing on Rezaei et al. (2017)

| Color | Thickness (mm) | VLT | SHGC | U ($W/m^2K$) |
|---|---|---|---|---|
| Bronze | 4 | 0.60 | 0.68 | 5.8 |
| | 6 | 0.49 | 0.60 | 5.7 |
| | 8 | 0.40 | 0.53 | 5.6 |
| | 10 | 0.33 | 0.48 | 5.6 |
| Green | 4 | 0.79 | 0.63 | 5.8 |
| | 6 | 0.73 | 0.55 | 5.7 |
| | 8 | 0.68 | 0.50 | 5.6 |
| | 10 | 0.63 | 0.46 | 5.6 |
| Gray | 4 | 0.55 | 0.66 | 5.8 |
| | 6 | 0.43 | 0.57 | 5.7 |
| | 8 | 0.34 | 0.50 | 5.6 |
| | 10 | 0.26 | 0.45 | 5.6 |

### 6.1.2 *Reflective Glass*

Reflective glasses are obtained through the application of metallic oxides in the form of coating (placed on the outer glazing of the IGU) that increases pane's reflection toward near-infrared radiations, reducing the transmission through them without shading the infrared waves (Fig. 6.3).

Also called *solar control glasses*, they are characterized by a predetermined selective reflection toward solar radiation and by a low SHGC. Compared to tinted glazing, reflective glass has a much greater impact on the reduction of solar transmission.

To obtain them, different types of surface deposits can be made, with various deposition techniques, such as additives methods, subtractive methods, cathode sputtering or vacuum evaporation, thermal spraying or pyrolysis, and chemical deposition. The most used methods are *pyrolysis* and *sputtering* which generate different performance glazing. The first deposition technique consists of the deposition of metallic coatings through a vitrification process at very high temperatures (about 600–650 °C); this process takes place continuously on the production line during the annealing phase of glass; therefore, it is rather inexpensive and guarantees an excellent fastening of the oxides to the support, giving a high resistance toward external atmospheric agents to the coating. On the other hand, the number of usable oxides is limited, and the production line does not guarantee a constant thickness of the deposit.

Instead, products obtained through magnetron sputtering deposition technique[3] (also defined as cathodic pulverization process) are treated after the completion of the base glass production, thus having higher costs but allowing a greater range of colors than the previous glasses, better homogeneity of coatings, and greater optical

[3] Process carried out under high vacuum in an electromagnetic field.

uniformity, although they are more delicate and the layer must be applied on face two or three to prevent the aggressive action of the atmospheric agents that can damage the metallic deposit (Fig. 6.4).

In general, the reflectivity of such systems increases when the thickness of the film increases, while light transmission decreases.

The outer installation of surface-treated glass allows to obtain glazed products with high lighting reflection values in the visible range (RL about 35/40%) and low in the infrared range; this feature, combined with a fairly high solar factor, recommends the use of reflective glass in climate contexts in which reduction of solar gains is fundamental (cooling dominated climates).

However, it has to be underlined that these types of glass allow a one-way view, which means visual transparency only from the less enlightened space toward the brighter one (Figs. 6.5 and 6.6).

### 6.1.3 Selective Glass

Selective glass employs coatings made of metallic oxides and dielectric materials, generally applied on the outer glazing, to block incoming infrared radiation, into indoor environments. They maintain discrete light transmission levels and limiting, at the same time, the solar factor's value. The selective coating has the function of filtering the electromagnetic waves, admitting most of the incoming solar radiation in both visible and near-infrared spectra; instead, long-wave radiations (far-infrared), emitted from warm objects within rooms, are reflected into the room. In this way, during the winter season, the inclination of the solar rays—on average parallel to the thin glazed element—make radiations able to penetrate through the system, thus triggering the greenhouse effect exploited in passive thermal-energy strategies to reduce heating loads.

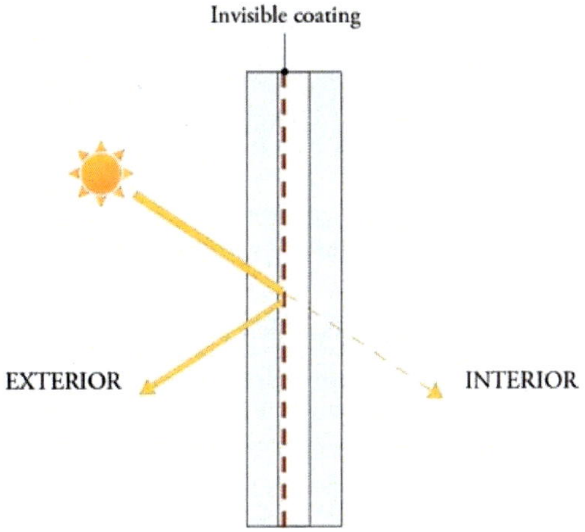


**Fig. 6.3** Operating principles of a reflective glazing system in a double insulated unit solution

Fig. 6.4 Pilkington Reflite™ Clear application. Courtesy of Pilkington Poland ©

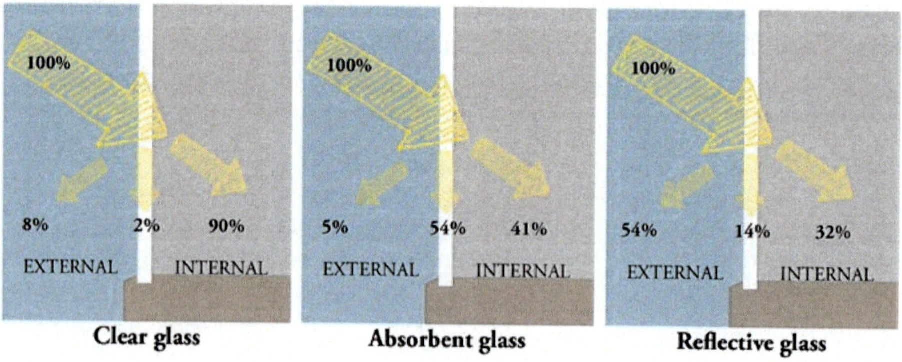


**Fig. 6.5** Different percentage of incident solar radiation entering through a glass surface, based on the type of glazed technology installed

These systems couple thermal insulation characteristics, typical of low-emission glass, with antisolar properties of reflective glass.

The selective ability they provide is achievable through the application of selective films to a common window pane, typically in the form of a separate sheet, that can be inserted between two or more glass sheets (Fig. 6.7).

A variation in the traditional selective glass is *angular selective glass*, traditionally equipped with a series of very small louvers in the air gap between the two glass

**Fig. 6.6** Use of both tinted and reflective glasses in a commercial building in Belgrade, Serbia. Photo credit: the author

**Fig. 6.7** Operating principles of a selective glazing system in a double-insulated unit solution

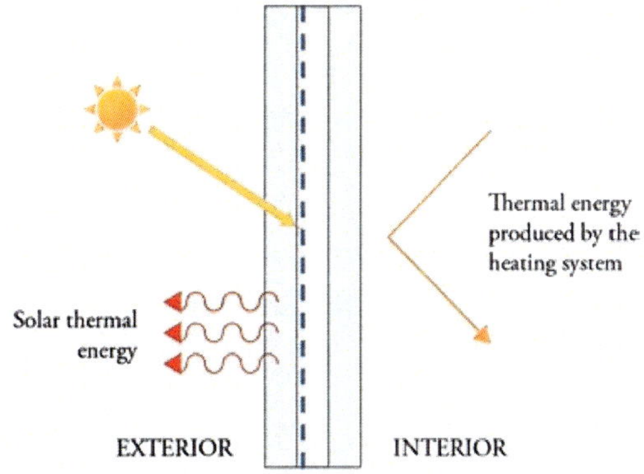


panes (Fig. 6.8). A novelty, recently introduced, is an angular selective device that embeds in the air gap a special honeycomb insert; this structure, combined with a proprietary polymer composition, provides the advantage of dynamic performance—increasing solar heat protection at peak hours—without the invisible light reduction that generally occurs with other dynamic technologies.

**Fig. 6.8** On the left: (**a**) OKASOLAR F product from OKALUX Glastechnik GmbH: a glazing system with louvers inserted in the cavity to direct the daylight selectively. Courtesy of OKALUX Glastechnik GmbH. On the right (**b**): application of OKASOLAR F in the office building "Profilmetall" in Marktheidenfeld—Photographer: Christian Schwab on behalf of OKALUX. Courtesy of OKALUX Glastechnik GmbH

**Fig. 6.9** Operating principles of a Low-Emittance system in double-glazing solution

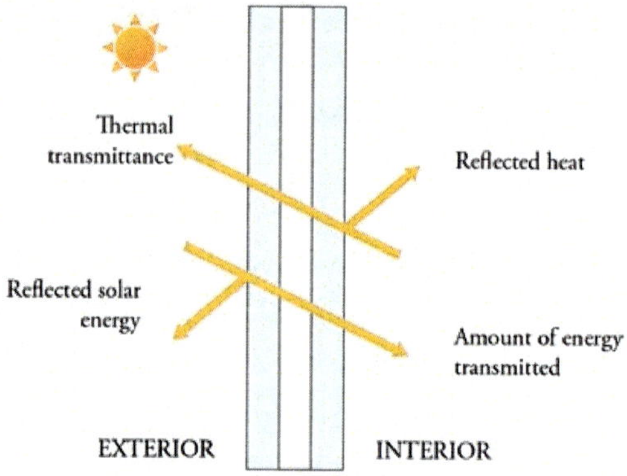


### *6.1.4 Low-Emissive Glass*

In Low-Emissive glass, a low-emissive treatment able to reduce heat exchanges through the system without compromising transparency is applied. These devices are transparent toward solar thermal radiations (capable of penetrating inside thermal environments) but reflective regarding infrared radiations; this results in an increase in the greenhouse effect due to its capacity in preventing the loss of the radiation emitted by the internal heating elements and allowing in this way its reflection toward the indoor spaces with consequent savings in energy costs for air conditioning (Fig. 6.9).

These systems are generally composed of a double-glazing unit filled with dried air, or with a noble gas, made reflective through the deposition of metals or semi-conductive metal salts, obtained by pyrolysis or by sputtering. This feature is also

**Fig. 6.10** Pilkington Optitherm application. Courtesy of Pilkington Poland ©

obtained through the insertion of an air gap between the two panes of a plastic film onto which a thin deposit of metal oxides is applied.[4]

A weak point of these systems is multiplying the effect of thermal solar radiation, which increase overheating; for this reason, they are particularly suitable to be applied in heat-dominate climates (rigid climatic contexts), where high solar factor (SHGC) is required as well as a low thermal transmittance coefficient (U-factor).

Low-E devices have total optical transparency and high light- and energy-transmission capacity (Fig. 6.10). However, the delicacy of metal oxide deposits implies that they have to be installed in double-glazing insulation systems, with the treated side toward the inside due to its sensitivity toward moisture.

The best effectiveness of the system is ensured by sandwiching the film between thin dielectric layers, able to reduce the reflectance of the coated glass in the solar spectral range.

Today, Low-E coatings are used in all major IGUs; depending on the thickness of the cavity and on the filling gas type, they allow to obtain thermal transmittance values between 1.7 and 1.0 W/m$^2$K, for double glazing, and lower than 0.7 W/m$^2$K for triple glazing. Liu (2012) shows that with the adoption of these technologies, the solar heat gain through windows can be reduced up to 48%.

[4]To provide Low-E coatings, two approaches are possible: *hard-coatings*, durable, less expensive but generally also less effective, and *soft-coatings*, with better thermal performance but more expensive and subject to oxidation during the manufacturing stage.

### 6.1.5 Vacuum Insulated Glass (VIG)

Introduced at the beginning of the century (Zoller 1924), Vacuum Insulated Glazing (VIG) was not produced until 1989; today, they are almost fully available on the market, even if, at quite a high cost.

Standard VIG products consist of two glass panes, generally low-emissive, separated by a vacuum space in which supporting and separating pillars stand to avoid the collapse of the system. The pillars, made of indium and disposed to properly react to atmospheric compressive stresses imposed by the pressure within the two panes (of about $10^{-2}$ mbar), have a very small dimension, so they have no impact on the vision toward the outside.

The advantage of such technology is that it combines an extremely small thickness with very good qualities of thermal insulation due to the vacuum between the glass panes that avoid convective exchanges between them; also, radiative exchanges are much lower, as the presence of selective coatings reduces thermal losses.

In the past years, several authors carried out experimental research to evaluate the thermal performance of vacuum glazing as accurately reported by Cuce (2016); last developments in this field are pointed toward sealants and spacers with high performances to increase the internal pressure of these systems to reach for a U-value of about 0.4 $W/m^2K$.

However, such a kind of window assembly still presents many shortcomings, for instance, the resistance to the variable pressures caused by wind and vibration that can affect the flat surface of windows, or, again, the maintenance of an airtight seal around the edge to avoid the re-establish of conduction to its normal value.

### 6.1.6 TIM Glass

Discovered in Germany in the 1980s (Tucci 2006), TIM glasses can be described as plastic or glazed components that comprise, in the space between the two panes, a Transparent or Translucent Insulating Material (Al-Obaidi et al. 2014). Their main advantage is that they combine diffused lighting features with very limited thermal losses that usually characterize glass curtain walls (Qu et al. 2014).

TIM glazed systems allow to reduce energy flows through the casing and, being translucent (SHGC is equal to that of a common IGU), they permit the diffusion of natural light—both direct as diffuse—in environments where there is no need to ensure the complete visual transparency toward the outside, preventing glare phenomena.

These systems exploit TIMs' main characteristic—that is their particular geometrical structure—to reduce radiative and convective heat exchanges through them.

Therefore, based on their geometrical structure, Transparent Insulation Materials can be divided into (a) *multiple parallel structured materials*, in which the insulation capacity of the material increases with the number of the elements and the internal reflections (reducing in this way also the solar radiation transmission) and (b) *glazed upright structured materials,* able to direct radiation through the inside.

**Fig. 6.11** On the left (**a**): KAPILUX insulating glass. In this panel, the translucent capillary tubes placed within the cavity divide incident solar radiation, allowing diffuse lighting in confined spaces. Glazed transmission of the system can be chosen specifically according to different needs. The small cavities between capillaries produce a sort of insulating air buffer able to stop the heat convection; for this reason, increasing the capillary tubes within the panels improves its thermal transmittance. In general, a system of such a kind offers an excellent insulation level with Ug value lower than 0.7–0.8 W/m²K. Courtesy of OKALUX Glastechnik GmbH. On the right (**b**): Application of KAPILUX panels in the Chloé Store in London. Photographer: Ali Moshiri on behalf of OKALUX. Courtesy of OKALUX Glastechnik GmbH

**Fig. 6.12** On the left (**a**): OKALUX light-diffusing insulating glass with capillary slab into the cavity. It provides very good sun and glare protection and excellent heat insulation. Thanks to a functional layer in combination with a gas-filled cavity for double- and triple-glazing build ups, an optimized Ug-value can be achieved. Courtesy of OKALUX Glastechnik GmbH. On the right (**b**): Application of OKALUX panels in the sports hall of Traunreut. Photographer: Ali Moshiri on behalf of OKALUX. Courtesy of OKALUX Glastechnik GmbH

To this group belongs *capillary structures* with parallel fibers and *honeycomb structures* with perpendicular fibers. Both honeycomb and parallel-fiber panels have to be applied with the structure direction perpendicular to the facade plane (parallel to the direction of incident solar radiation) to limit the number of internal reflections through structures (Figs. 6.11 and 6.12).

Glazing TIM systems generally employ plastic capillaries or honeycomb structure insulating materials that can be made of polycarbonate or poly-methyl-methacrylate (honeycomb structured compound of organic nature), equipped with excellent light-transmission and heat-insulation properties (90% for normal incidence), or an Aerogel, an inorganic material (with porous and hygroscopic structure), made of more than 90% of air and, for the remaining percentage, of

**Fig. 6.13** Different forms of Aerogels: (**a**) (on the left) Aerogel in monolith tiles. Source: https://commons.wikimedia.org/wiki/File:Aerogel_1_2016-05-16.jpg by FASTILY (own work). CC BY-SA 4.0 (https://creativecommons.org/licenses/by-sa/4.0) via Wikimedia Commons. (**b**) (on the right) Aerogel in granules. Source: https://commons.wikimedia.org/wiki/File:Aerogel_Vial.jpg by Zach (own work). CC BY-SA 2.0 (https://creativecommons.org/licenses/by-sa/2.0) via Wikimedia Commons

silica-dioxide particles, with very low thermal conductivity (lower than 0.013 W/$m^2$K) (Baetens et al. 2011).

Aerogel, due to the high percentage of air contained in it, presents very low density and very interesting thermal characteristics; extremely fragile and very moisture-sensitive, it is available in different forms: the most common forms are monolithic and granular forms (Buratti and Moretti 2011) (Fig. 6.13). Monolithic aerogels are panes of thickness between 8 and 20 mm, with a density between 3 and 500 kg/mc; they can be directly used to replace traditional glazing even if their high cost, coupled with the small available size of the panes, makes them difficult for application. Glazed systems produced with this material—appeared on the market in 2005—have a high solar transmission coefficient (up to 80%) and a very low heat transmission coefficient; for these reasons, they are suitable to be used both in cooling- and in heat-dominant climates.

Aerogel in granular form instead, made of spheres with variable diameter, can be easily used in IGUs even if it has a very low light transmittance.

In general, these kinds of devices allow obtaining very thin window systems, with U-value <0.50 W/$m^2$K and without diminishing the solar factor nor considerably reducing VLT.

### 6.1.7 *Self-Cleaning (SC) Glass*

Self-cleaning glass exploits technologies and formulations able to transform the surficial properties of glass by penetrating the micro-pores of the glazed surfaces. These applications have the advantages to reduce maintenance needs while controlling solar radiation by diminishing depositions on the glass surface.

Two categories of self-cleaning treatments can be identified: *hydrophobic* (water repellent) and *hydrophilic* (water attractive) surfaces. A solid surface with a contact angle (between itself and a water droplet) higher than 90° is considered hydrophobic, whereas a contact angle lower than 90° makes it hydrophilic. If this angle exceeds 120°, the surface is considered superhydrophobic (also defined Easy-To-Clean (ETC) surface); similarly, when the angle is lower than 60°, the surface is called superhydrophilic.

Self-cleaning hydrophobic surfaces have surficial treatments of nano-technological t, capable of reacting with silica's molecular chain of glass, thus eliminating its micro-porosity (Mellot et al. 2006). These formulations transform glass' surface properties, uniformly arranging molecules and thus developing a surface layer capable of letting water and surface deposits flow in a perfectly smooth and compact way and in contrast with the natural surface of common glass, which has, on the contrary, imperfections that led it to accumulate water, corrosive substances, or dirt (Fig. 6.14). These surfaces are also defined as *lotus effect* surfaces (Midtdal and Jelle 2013) and, thanks to the reduction of surface adhesion, can reduce cleaning requirements, obtaining water repellent systems in which contamination is washed off by water droplets (Fig. 6.15).

On the other hand, hydrophilic glasses are generally obtained through the deposition of metallic oxide layers (generally $TiO_2$, ZnO, or $WO_3$) on their surfaces; commercial manufacturing products are mostly made with atmospheric pressure chemical vapor deposition techniques (APCVD) or with other various techniques, such as PVC, CVD, sol-gel, electrospinning, and dip-coating.

The treated surfaces, due to the photocatalytic properties of the metal-oxide-based nanoparticles, decompose the organic contamination using UV radiation. The coating,

**Fig. 6.14** Water droplets beaded up on a hydrophobic surface. Source: https://commons.wikimedia.org/wiki/File:What_a_little_Rain-X_can_do_for_you_(3908431451).jpg by Michael Mol (own work). CC BY 2.0 (https://creativecommons.org/licenses/by/2.0) via Wikimedia Commons

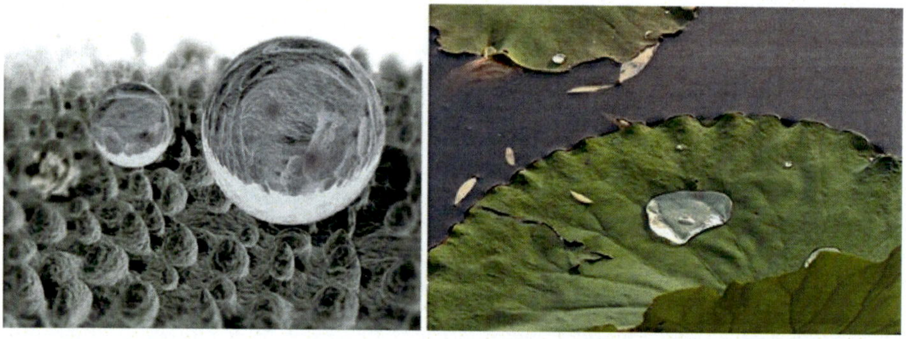

**Fig. 6.15** Schematic functioning of the so-called lotus effect. On the left (**a**): microscopic image of a lotus leaf with water droplets on its surface. Source: https://commons.wikimedia.org/wiki/File:Lotus2mq.jpg by William Thielickewebsite (own work). CC BY-SA 3.0 (http://creativecommons.org/licenses/by-sa/3.0/) via Wikimedia Commons. On the right (**b**): water droplets on a lotus leaf. Source: https://commons.wikimedia.org/wiki/File:Lotus_effect_(7894611634).jpg by Nikolaj Potanin (own work). CC BY-SA 2.0 (https://creativecommons.org/licenses/by-sa/2.0) via Wikimedia Commons

interacting with UV rays through a photocatalytic process, provides the decay of organic pollutants and nitrogen oxides, allowing rainwater to wash away dirt. The photocatalytic properties long last until the irradiation continues, while after illumination stops, the hydrophilic behavior remains for about 2 days (Fig. 6.16).

Nowadays, super-hydrophilic devices are also available which employ nanoscale polymers or silica nanoparticles to provide a perfectly smooth surface able to perform an antifogging function.

The difference between photocatalytic self-cleaning surfaces and ETC systems is that ETCs do not require UV light to function and that their hydrophobic properties—on the contrary of hydrophilic devices—determine the runoff of water droplets rather than form a hydrophilic film on them. Both these kinds of glazing are particularly suitable for difficult-to-reach surfaces.

Thanks to the reduction of dirt on their surfaces, self-cleaning systems improve light transition through them, reducing energy costs for lighting accordingly.

The ratio of decontamination to contamination expresses the efficiency of the glazing.

## *6.1.8 Heating Glass*

Heating glasses represent a pretty promising smart window technology, since they seem able to contain heat and gain losses through the transparent surface by adding to the interior pane of an IGU a metallic coating, electrically connected (generally made of aluminum oxide). When a low voltage is applied, the glazing warms up and directs the heat to the interior, avoiding heat exchanges through the room and the external environment.

Systems of such a kind completely remove the risk of condensation between glass panes, contributing to the heating system of the entire building (since they generate about 0.42 $kW/m^2$).

**Fig. 6.16** Sunlight action on a photocatalytic glass

Normally, heating glasses require 100–300 W/m$^2$ to work, providing a pane temperature of about 40 °C; if used only for condensation-prevention purposes, their operation power can be halved.

Further improvements of such technology include the development of new high-conductive oxide coatings with nanoparticle for faster heating rates, lower consumption, and higher color neutrality, which means increased VLT.

Appeared on the market only in 1980, and first in Finland, now heating glasses are produced by several companies all over the world and at a quite low cost when compared to other smart window technologies.

### *6.1.9 Photovoltaic (PV) Glass*

Photovoltaic glasses embed photovoltaic cells or panels into the system to transform solar energy into electrical energy, acting in the same way as a sunscreen.

To the author's knowledge, several types of products are available on the market, using different photovoltaic elements, although silicon (Si) still remains the prevalent semiconductor used in solar cell production. It is used in the form of crystalline silicon, mono- or poly-crystalline, or non-crystalline cells, produced by amorphous silicon; noncrystalline cells have low average efficiency (around 9%) even if they are available in both opaque and semi-transparent modules. Otherwise, monocrystalline cells are blue-, dark gray-, or black-colored opaque elements with a degree of efficiency of about 15% (Jelle et al. 2012a, b). Finally, polycrystalline cells are blue opaque and made with silicon-drained blocks and are less efficient

**Fig. 6.17** Polycrystalline solar cells at Tübingen's sports hall. Source: https://commons.wikimedia.org/wiki/File:Photovoltaic-cells.jpg by Björn Appel (own work). CC BY-SA 3.0 (http://creativecommons.org/licenses/by-sa/3.0/) via Wikimedia Commons

(their degree of efficiency is about 12%); they are produced with pane thickness of 0.4 mm and sizes of 10 × 10 cm or 15 × 15 cm, combined into modules and placed in IGUs' airgap using resin (Fig. 6.17).

Other PVs glasses are those produced using thin-film organic materials—semiconductors, cheaper than inorganics, like Si, but still sensitive concerning degradation—or dye-sensitized solar cells (DSSCs) covered by a thin film and constituted by a photo-sensitized anode and an electrolyte. DSSCs presents low cost but very poor energy conversion efficiency (Cuce 2016) (Fig. 6.18; Table 6.2).

To overcome the major weak point of these systems (that is the lack of visible transmittance), researchers of the Michigan State University developed a PV technology that uses a particular kind of plastic to produce electricity, if directly irradiated (Peschel 2015). It is called Transparent Luminescent Solar Collector, and it can be placed over windows to absorb specific nonvisible wavelengths.

This new harvesting system has been developed using organic molecules that absorb UV and NI wavelengths and, transporting thin strips of photovoltaic cells on plastic, convert light into electricity. Since absorbed wavelengths are not visible to the human eye, the material looks transparent, opening new interesting possibilities for the world of Windows' Integrated Photovoltaics (WIPVs). However, despite its many upsides, the real shortcoming of this technology is its low energy efficiency, still around 1%.

**Fig. 6.18** The EPFL QUARTIER NORD SwissTech Convention Center Retail and Student Housing (Richter Dahl Rocha & Associés, 2014, Lausanne, Switzerland) has sunscreens made with DSSCs device. Source: https://commons.wikimedia.org/wiki/File:SwissTech_Convention_Center_02.jpg by Ramoul (own work). CC BY-SA 3.0 (https://creativecommons.org/licenses/by-sa/3.0) via Wikimedia Commons

**Table 6.2** Different specifications for six semi-transparent PV modules. Author's editing on Rezaei et al. (2017)

| Dimension | Features | VLT | SHGC | U (W/m$^2$K) |
|---|---|---|---|---|
| 980 × 950 mm | Efficiency \| 8.02%<br>PV technology \| a-Si<br>Assembly \| Single pane laminated | 0.0917 | 0.289 | 5.08 |
| 1300 × 1100 mm | Efficiency \| 5.90%<br>PV technology \| μc – Si<br>Assembly \| Single pane laminated | 0.0519 | 0.413 | 4.80 |
| 1300 × 1100 mm | Efficiency \| 3.32%<br>PV technology \| μc – Si<br>Assembly \| Single pane laminated | 0.0184 | 0.298 | 5.08 |
| 1300 × 1100 mm | Efficiency \| 4.43%<br>PV technology \| μc–Si<br>Assembly \| Single pane laminated | 0.0417 | 0.387 | 5.10 |
| 989 × 930 mm | Efficiency \| 5.01%<br>PV technology \| a-Si<br>Assembly \| Double pane unit | 0.0691 | 0.154 | 1.67 |
| 980 × 950 mm | Efficiency \| 4.75%<br>PV technology \| a-Si<br>Assembly \| Double pane unit | 0.0734 | 0.123 | 2.14 |

Also, an Italian research team tried to develop the first transparent photovoltaic glass with the introduction of graphene, a material tougher than diamonds, totally transparent, able to conduct the electricity even if employed in thickness comparable to that of an atom and with considerably reduced costs, when compared to traditional materials.

In the study (Giucastro 2017), the platinum that normally presents in a PV panel (based on dye-sensitized solar cell (DSSC) technology) has been entirely replaced by graphene, in the form of ink, and applied to the panel by spray deposition.

The advantage of this technology is that it can further be implemented with higher energy performance and that the production techniques of graphene panels could easily be produced industrially.

## 6.2 Active | Dynamic Glazing Technologies

Suitable to be inserted within the category of *chromogenics*, glazing systems of such a kind are essentially characterized by properties of variable transmittance. Such ability is obtained through the integration, within the system, of chromogenic materials or of materials capable of changing their optical properties, according to external stimuli or after the application of an electric field. Chromogenics, indeed, exploits chemical–physical properties of some technologies to change their status; in the case of glass, they change it from highly transparent to partially reflecting. Even if, in the first state, the glass pane allows high visible transmittance and Solar Heat Gain Coefficient, in the colored state, these two values are expected to be as low as possible.

According to the technologies employed, thus of their behavior, chromogenic glass can be self-regulating (thermal and photochromic) or electronically activated (on the basis, for instance, of users' demands) (electro-chromic).

To the author's knowledge, chromic-materials-based smart windows are the most reliable technologies for producing smart adaptive glazing systems; however, even if such technologies are almost fully exploited, they still do not provide full control of direct sunlight, since they directly turn from a transparent state to a tinted one, without intermediate transparency levels.

### *6.2.1 PCM Glass*

PCM systems embed Phase Change Materials to control incoming solar loads, decreasing building energy loads in peak hours and impeding building temperature swing.[5] We can say that such kind of systems integrates four functions in a single

[5] In a study (Nghana and Tariku 2016) it was found that wall and indoor air temperature fluctuation is decreased by 2.7 °C and 1.4 °C respectively in a building that incorporates PCM; moreover, also energy demand was reduced by 57% during winter.

functional unit, which can be resumed as follows: (1) the insulating function, (2) the overheating control function, (3) the energy conversion function, and the (4) storage function, thanks to the presence of the PCM.

Different materials are suitable to be employed as PCMs, exploiting the principle of latent heat thermal storage (LHTS) (to absorb energy and release it at a different time) and changing their state from solid to liquid, allowing in this way the temperature control in a specific range and function of the material used.

PCM glasses generally allow diffused natural lighting of spaces; at the solid state indeed, PCMs transmit about 28% of visible light, while, when they are in a liquid state, VLT increases to more than 40% (Goia 2012).

Although translucent PCMs are available (Long et al. 2014), the light quality transmittance of these technologies is still a weak point. Moreover, PCMs still have obvious limitations that need to be overcome to reach large-scale application, such as, for instance, issues related to the choice of the optimal melting temperature, flammability of paraffin, and the difficulty to thermally download the material after long "warm periods."

### *6.2.2 Suspended Particle Devices (SPDs) | Liquid Crystal Devices (LCDs)*

Suspended Particle Devices (SPDs) are electroactive devices in which the application of an AC voltage makes the particles move from a random pattern to an align-oriented one, making the window transparent. Without an electrical charge, SP windows absorb light and thus decrease light transmission through them.

Generally, in SPDs, transparency depends on fluid film thickness and particle density; in addition, they can be tuned at different levels quite quickly even if with pretty high energy consumption.

Typical ranges of VLT and SHGC in these devices—in the transparent and opaque state—are, respectively, for VLT: 60–0.5%, and, for SHGC: 0.57–0.06%, with switching times of some seconds. They require about 100 V AC to operate from the tinted state to the transparent state and can be modulated via intermediate states. Power requirements are about 5 $W/m^2$ for switching and 0.55 $W/m^2$ for maintaining constant transmission (Table 6.3).

On the other hand, Liquid Crystal Devices (LCDs) use materials with a bar-molecular structure that, when a voltage is applied, allows to vary systems' light transmission properties; most of the LCDs indeed, scatter light, become white and translucent (Rezaei et al. 2017) (Fig. 6.19). These materials, discovered by Gray in Great Britain during the 1970s, are defined as *nematic* if they are align-oriented but can shift each other, or *smettic* if they are stratified.

Different kinds of LCDs become smettic at low temperatures, while nematic if heated, and usually fluid when they reach a very high temperature. Most common LCDs use crossed-nematic devices, whose polymers form chains that round between polarized plates.

LCDs are generally composed of a layer of droplets dispersed in a polymer matrix; the particles are inserted between two transparent electrodes made of

**Fig. 6.19** LCD glass switched on, employed for privacy purposes. Source: https://commons.wikimedia.org/wiki/File:Shtiever_2.jpg by Shtiever (own work). CC BY-SA 3.0 (https://creativecommons.org/licenses/by-sa/3.0) via Wikimedia Commons

**Table 6.3** SP windows' specifications. Author's editing on Rezaei et al. (2017)

| Performance | OFF state | ON state |
|---|---|---|
| VLT | 0.65 | 0.01 |
| SHGC | – | – |
| U (W/m$^2$K) | – | – |
| Maximum power draw—19.4 mA/m$^2$; Switching time—1–3 s | | |

**Table 6.4** LC privacy window's specifications. Author's editing on Rezaei et al. (2017)

| Performance | OFF state | ON state |
|---|---|---|
| VLT | 0.77 | 0.58 |
| U (W/m$^2$K) | 1.59 | 1.59 |
| Operating voltage (AC) | 120 V | |
| Power consumption | 15.5 W/m$^2$ | |
| Switching time | 0.1–0.4 s | |

ITO-covered polyester. When the voltage is off, they scatter light due to the mismatch between the matrix and droplets' refractive index.

These systems, unlike SPDs, are mainly employed for privacy purposes (Kim et al. 2015) and require current electrical energy to maintain operation.

The light transmittance of liquid crystal glazing, in the active state, does not exceed 70%, while in the off state it is about 50%; they spread direct incident solar radiation but do not block it enough to reduce SHGC, which lays generally between 0.69 and 0.55 (Table 6.4).

### 6.2.3 OLEDs

Slightly different than the other glazed technologies analyzed so far, the OLED glass technology fits in this category due to its ability to change properties of glass over time if an electrical input is applied.

Indeed, Organic Light Emitting Diodes' technology consists of two glass sheets on which, during the production phase, thin layers of hydrocarbon-based chemicals are vapor deposited. Unlike conventional LEDs and other light sources, OLEDs can emit their light across the entire surface; this feature makes them the first real flat area light source.

Since their light is roughly comparable to the natural skylight and glare-free, and that they are very thin (measuring approximately between 1 and 2 mm), OLEDs are suitable for materials so far not appropriate for lighting applications, such as glass.

These devices are mirror-like in the switched state when a layer of aluminum is used as a cathode, while when it is replaced by silver, they appear transparent; so, it is clear that the possibility of emitting light from a seemingly transparent glass (without a visible light source when switched off) makes this technology a unique opportunity.

Even if engineers are working on special building applications of these technologies, many of them—unfortunately—have not been developed beyond the prototype stage; this means they are not yet market-ready because of too high material and installation costs, coupled with too short service life.

### 6.2.4 Electrochromic (EC) Glass

Part of the so-called chromogenic devices (Granqvist 2006; Granqvist et al. 2009, 2010; Tavares et al. 2014; Carmody et al. 2014), electrochromic glasses are systems able to vary their light transmission properties (generally by varying their color and optical features) after the application of an electric field.

Electrochromics are obtained by interposing in the gap between two glass panes a layer of micro-liquid crystals[6] capable of ensuring reversible electrolytic reactions that, if subjected to a potential difference, can vary their coloration up to becoming transparent (Gorgolis and Karamanis 2016).

EC systems are based on both organic and inorganic materials. Organic ECs are made of molecules that change their color during oxidation/reduction processes, while inorganic ECs are based on transition metal oxides (mostly W and Ni). Two

[6] Micro-liquid crystals are metals generally based on indium and silver oxide, $WO_3$, or $MoO_3$ Molybdenum Tungsten. The surface layer is applied with vacuum and physical deposit of vapors techniques (PVD).

types of metal oxides exist (cathodic and anodic), and they both are simultaneously employed in standard EC devices.

In EC systems, ions move through an electrolyte layer due to an external electric field induced between two transparent conductions (from one electrode to another); through this migration, a new state of aggregation is obtained, favoring a new glass color and modifying light transmission of the system from about 1–4%, in the opaque state to 60–63% in the transparent state; the SHGC instead stays between 0.06 and 0.46.

If, in the initial stage, crystals have a milky-white color—obstructing light transmission through ray light deflection in different directions (through molecular disorder)—after the application of an electric charge, crystals will be oriented differently, thus modulating the system's optical behavior.

Once the change has occurred, no electricity is needed for maintaining the shade that has been reached. Even the amount of energy required to activate these devices is minimum.

The switching time of EC devices varies from seconds to minutes depending on the technologies employed and on window size.

Most of today's available devices operate in on- or off-states only, although technologies that allow variable levels of transparency are easily applicable (Tavares et al. 2014).

Recent advancements in electrochromic materials concerning transition-metal hydride electrochromic have led to the development of reflective hybrid systems, which become reflective rather than absorbent and thus which switch state between transparent and mirror-like. These materials are based on the same concept of traditional electrochromic even if they go about the problem in a different way, that is, by switching between a transparent state, when they are off, to a reflective state, when a voltage is applied (Fig. 6.20).

Additionally, a special type of electrochromic window has been developed at the Cockrell School of Engineering of the University of Austin, Texas (Llordés et al. 2016). This smart system employs two active electrochromic materials to choose between three different operation settings: one scenario admits the entry of light and solar radiation; another scenario blocks only infrared radiations while the last scenario prevents entry of both light and heat. This is possible, thanks to a double-band electrochromic material that incorporates elements for light-selective control, with different optical properties in a very low switching time.

Also, self-powered EC windows that employ PVs for being activated are under development (Cannavale et al. 2015) even if their transparency is very low due to the presence of the PV layer. These devices employ sputtered titanium and tungsten oxide films as an electrochromic layer, coupled with a photoactive layer made of dye solar cells (generally made of dye-sensitized titanium oxide—$TiO_2$) (Bogati et al. 2017).

Photo-electrochromic devices can provide a twofold behavior, being able to generate energy and, at the same time, to switch their transparency and light-transmission mode (Cannavale et al. 2016).

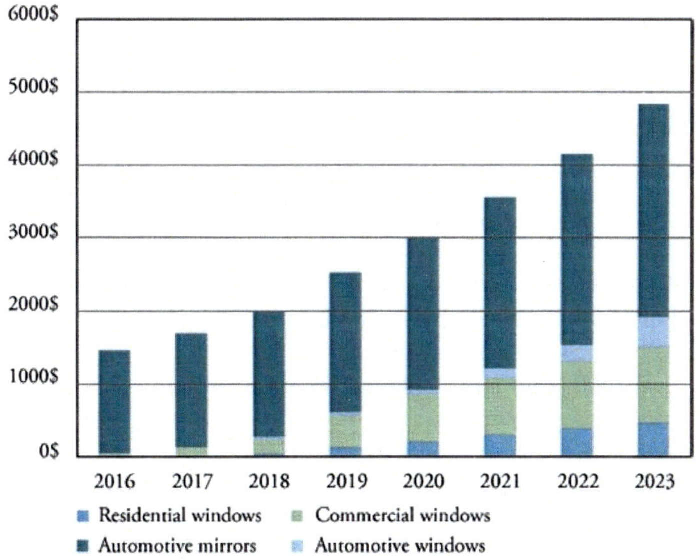


**Fig. 6.20** Forecasts for Electrochromic glass markets in terms of millions of dollars. Author's editing on n-tech Research data (2016)

**Table 6.5** Specifications for a commercial EC window. Author's editing on Rezaei et al. (2017)

| Performance | OFF state | ON state |
|---|---|---|
| VLT | 0.60 | 0.01 |
| SHGC | 0.41 | 0.09 |
| U (W/m$^2$K) | 1.59 | 1.59 |

Despite their increasing development and appeal, these devices still lack visual transmittance (still about 61.5%—in the bleached state—and 11.2% in the colored state) (Table 6.5).

## 6.2.5 *Thermochromic (TC) Glass*

Thermocromics are glazed systems in which a temperature variation triggers a reaction in the material, giving it a high reflective capacity thus making it more sensitive toward infrared radiation. This results in a light radiation absorption variation as a function of the external surface temperature, making these devices opaque when a critical temperature (specific for each product) is reached.

The alteration of optical properties is obtained through the modification of the molecular balance, an induced chemical reaction, or the transition of the crystalline structure.

In general, the most employed TC technology is the one that uses Tungsten trioxide or Vanadium dioxide coatings to obtain this totally reversible behavior (Hee et al. 2015). Vanadium dioxide ($VO_2$) is the most common material, despite its many shortcomings, such as the high transition temperature, comprised between 10 °C and 65 °C, so much higher than indoor temperatures (Li et al. 2012). To overcome these drawbacks, other materials, such as W and F, are added to $VO_2$ to decrease its transition temperature, or special gels are inserted between the two plastic film layers. However, as a result, the VLT of such devices is low; for this reason, antireflective coatings are used to increase it (Gorgolis and Karamanis 2016).

Critical points of thermochromic technology are the nongradual transition from transparency to opacity, as well as the vision that it produces, which seems to be not extremely clear. This feature makes thermochromics particularly suitable for applications in which visibility is not a priority, while it is desirable for privacy or a shielding effect.

To evaluate TC performance, solar modulation efficiency is used (Rezaei et al. 2017); it is denoted by $\Delta T_{sol}$ that expresses the difference between the solar energy transmitted before and after the transition. Such value ranges from 0 to 1 and, for acceptable performance, it should exceed 0.1 (Table 6.6).

To overcome shortcomings of existing thermochromic devices, scientists of the U.S. Department of Energy's National Renewable Energy Laboratory (NREL) developed a highly efficient window able to convert sunlight into electricity. Such a prototype employed perovskites and single-walled carbon nanotubes to respond to heat: During the transformation from transparent to tinted, the window darkens, generating electricity. The color change is driven by molecules (methylamine) that are reversibly absorbed into the device. When solar energy heats up the device, the molecules are driven out, so it is darkened, while when the sun is not shining, the device is cooled down, appearing transparent because the molecules reabsorb into it.

Such a device allows an average of 68% of VLT when it is in a transparent or bleached state, while when the window colors, only 3% is allowed.

The aim of the researchers was to develop a system able to perform a twofold behavior: from one side as a solar cell—when there is a lot of solar radiation—and from the other as a good window system when there is not. Results published in Nature Communications (Wheeler et al. 2017) showed a solar power conversion efficiency of the system of about 11.3%.

Such efforts were turned into the production field within the Emerging Technologies program funded by the Department of Energy's Building Technologies Office, under the name of SwitchGlaze.

**Table 6.6** Specifications for a thermochromic window. Author's editing on Rezaei et al. (2017)

| Performance | 25 °C | 65 °C |
|---|---|---|
| VLT | 0.50 | 0.12 |
| SHGC | 0.29 | 0.13 |
| U ($W/m^2K$) | 1.36 | 1.36 |

### 6.2.6 *Thermotropic (TT) Glass*

Thermotropic devices dynamically alter their solar and visible light transmittance and reflectance in response to the window's temperature (Yao and Zhu 2012). They differ from other types of chromogenic devices in two major aspects: they switch between a transparent and an opaque state, meaning that visibility is not retained in the "solar control" mode, and they react automatically to changes in the glazing temperature rather than being activated by an external switch. These two properties make this type of glazing particularly suitable for clerestory windows or skylights, where a direct view is not significant.

Their ability to change state is based on Lower Critical Solution Temperature (LCST) of a PCM (a two-component fluid[7]) inserted within the two glass' panes; up to a determined temperature (defined switching or transition temperature), the fluid remains in the form of a homogeneous, transparent mixture, while when it passes such temperature, it changes its configuration and, through the separation of the two components, makes the layer white-opaque, reflecting and diffusing much of the light, and thus decreasing the passage of light radiation. The switching process is reversible; if the outside temperature decreases below the switching temperature, the glass becomes clear again.

The advantage of using such technology is that during winter it remains clear even if the sun is shining, so daylight can be maximized at its best as well as energy demands for heating and lighting can be minimized accordingly.

Thermotropic materials can be divided into different systems based on the mechanism by which they achieve a state of low visible- and solar transmission, under the transition temperature. The three main groups are *thermotropic hydrogels*, *thermotropic polymer blends,* and *embedded thermotropic polymers* within fixed domains.

Recent research (Allen et al. 2017) tried to develop and test new types of thermotropic windows where the thermotropic (TT) layer, made of Hydroxyl-Propyl-Cellulose (HPC), is sandwiched as a membrane between two glass panes. Tests done on the HPC-based prototype show that in the transparent state (below 40 °C), it has VLT of 0.9 and solar transmittance of 0.75; while, above the transition temperature (over 50 °C), it has VLT of 0.14 and solar transmittance of 0.11.

In general, it can be said that TT devices allow obtaining transmittance above the 85% in the transparent state (under the transition temperature) and below 15% in the translucent state (above the transition temperature).

[7] The basic mixture interposed between the glazed panes is generally composed of two components with different reflection index (e.g., water and plastic—hydrogel—different plastics—polymer mixture).

### *6.2.7 Photochromic (PC) Glass*

Photochromic glass is a particular type of chromogenic device which originated approximately in 1899, after the discovery of the phenomenon of discoloration of some crystalline substances as a result of exposure to light radiation.

These devices are capable of varying their optical properties due to external light intensity variations, due to the presence of organic or inorganic compounds that act as optical sensitizers (Casini 2016). The photochromic reaction triggers an exchange between silver and copper ions, resulting in the precipitation of silver metal which determines the coloring of the glass pane.

PCs behavior is reversible when radiation stops, and this is possible, thanks to the decomposition of micro-crystals of silver halide (chlorides, bromides, and iodides) present in the glass paste and sensitivity to UV rays.

In such systems, light transmission depends on the amount of light on glass surface because they modify their transmission curve based on the intensity, the duration, and the nature of the incident solar radiation.

One of the limits of photochromic technology is the poor light transmission during the winter season, thus decreasing SHG in internal environments.

Since the photochromic glass coloring process is ambient temperature-sensitive—being more intense at low temperatures and essentially null at higher temperatures—it follows that their use in buildings is very poor (Granqvist et al. 2009).

### *6.2.8 Gasochromic (GC) Glass*

Devices obtained by the incorporation of a gasochromic (GC) layer between two panes of an Insulated Glass Unit (generally among the outer and the middle pane of a triple IGU), which, when exposed to diluted hydrogen gas (generally mixed with Ar gas), colors and changes system's transparency because of a catalytic reaction with the glass mixture (Wilson 1995). Their transparency level depends on the volume of hydrogen to which the gasochromic layer has been exposed. The second gas-filled cavity and the third pane, with low-emissive coating, ensure that IGUs of such a kind have an $U_g$ value of about 0.9 W/m$^2$K.

The most employed GC material is Tungsten oxide (Feng et al. 2016), coated with a thin catalyst layer, even if devices equipped with a thin layer of Wolfram (W) are available.[8] In systems with W, the contact with a low concentration of hydrogen is enough to confer to the panel a blue color; when the contact occurs with oxygen instead, the film bleaches to the original transparent state.

[8] The optically active component of a gasochromic IGU is a film of $WO_3$, thick less than 1 μm, which is coated with a thin film of a catalyst. It is located on the inner surface of the outer pane of a triple IGU.

The modification that occurs in glass transmission properties allows reducing visible and total solar energy transmittance values of a triple glazing unit from 0.63 and 0.49 to 0.20 and 0.17 respectively (when the indoor pane is coated with a standard Low-E coating). If a solar control coating is added, lower SHGC values can be obtained.

With GCs, clear visibility from inside to outside is retained in all switching states.

In contrast to conventional, external shading systems, gasochromic glazing can also be used in the upper floors of high-rise buildings even if they need more control equipment; in addition to a gasochromic insulating glazing unit, the system needs two further main components: a gas supply unit and a control unit. The gas supply unit consists of an electrolyzer and a pump, which is connected by pipes to the window in a closed-loop configuration. Ideally, the gas supply unit is integrated into the external building facade. One gas supply unit is able to provide sufficient gas to switch 10 $m^2$ of gasochromic glazing (Rezaei et al. 2017). The control unit instead permits both manual and automatic control. The integration into a bus system allows the glazing to be switched to optimize lighting conditions, thermal comfort, and/or building energy consumption (Fig. 6.21 and 6.22).

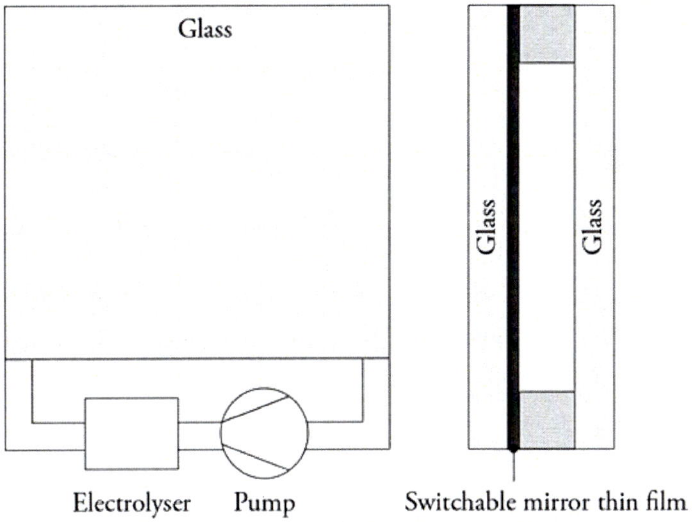


**Fig. 6.21** Conceptual scheme about the operation mode of a gasochromic system. Author's editing (Source: https://phys.org/news/2013-03-switchable-mirror-sheet-gasochromic-method.html)

**Status**

on-the-market available | under development | research

**Measurement unit**

U-value (W/m²K) | VLT (%) | SHGC (%) | LSG | Max. dim. (mm)

| | Clear float glass (6 mm thick) | Double IGU (4 \| 12 \| 4 mm) | Double IGU Argon (4 \| 16 \| 4 mm) | Triple IGU (4 \| 12 \| 4 \| 12 \| 4 mm) | Tinted glasses (IGU 6 \| 12 \| 6 mm) |
|---|---|---|---|---|---|
| U-value | 5,8 | 2,8 | 1,2 | 1,5 | 2,8 |
| VLT | 90 | 80 | 81 | 72 | 48 |
| SHGC | 87 | 70 | 73 | 50 | 41 |
| LSG | 1,03 | 1,14 | 1,10 | 1,44 | 1,17 |
| Max. dim. | 6000x3210 | 6000x3210 | 6000x3210 | 6000x3210 | n.a. |
| Cost | 16 €/m² | 50 €/m² | 65 €/m² | n.a. | 200 €/m² |

| | Reflective glasses (6 mm thick, single pane) | Selective glasses Argon (6 \| 16 \| 4 mm) | Low-E glasses (IGU 6 \| 12 \| 6 mm) | Vacuum insulated (Pilkington Spacia) | TIM glasses (IGU– 4 \| 20 \| 4 mm) |
|---|---|---|---|---|---|
| U-value | 5,7 | 1,0 | 1,9 | 1,4 | 0,5 |
| VLT | 31 | 67 | 53 | 76 | 70 |
| SHGC | 48 | 45 | 45 | 66 | 75 |
| LSG | 0,64 | 1,48 | 1,20 | 1,16 | 0,93 |
| Max. dim. | 6000x3210 | 6000x3210 | 3048x1778 | 2400x1350 | n.a. |
| Cost | 42 €/m² | 148 €/m² | 200 €/m² | n.a. | n.a. |

| | Self-cleaning glasses (Pilkington Active Clear) | Heating glasses (IGU 4 \| 16 \| 4 mm) | Photovoltaic glasses (IGU 4 \| 12 \| 4 mm) | PCM glasses (GLASSX Crystal) | LCD glasses (IGU 4 \| 12 \| 4 mm) |
|---|---|---|---|---|---|
| U-value | 2,6 | 1,1 | 1,67 | 0,48 | 1,59 |
| VLT | 75 | 74 | 6 | 50-80 | 77-58 |
| SHGC | 71 | 53 | 12 | n.a. | 60 |
| LSG | 1,06 | 1,39 | 2 | n.a. | ~1,12 |
| Max. dim. | n.a. | 3210x2200 | 1000x600 | 2800x150 | 900x2600 |
| Cost | n.a. | 110 €/m² | 340 €/m² | n.a. | 1000 €/m² |

| | Electrochromic glasses | Thermochromics glasses | Thermotrophic glasses | Photochromic glasses | Gasochromic glasses |
|---|---|---|---|---|---|
| U-value | 1,6 | 1,36 | 1,15 | 1,10 | n.a. |
| VLT | 58-1 | 50-12 | 72-23 | 25-62 | n.a. |
| SHGC | 41-9 | 29-13 | 48-15 | 22-27 | n.a. |
| LSG | 1,41-0,11 | 1,72-0,92 | 1,5 | 1,13-2,29 | n.a. |
| Max. dim. | 1828x3048 | n.a. | n.a. | 1000x2600 | n.a. |
| Cost | n.a. | n.a. | n.a. | 800 €/m² | n.a. |

**Fig. 6.22** Comparison among mostly of Passive and Active glazing technologies investigated; performance values are referred to technologies considered in general terms and not to specific commercial products, if not specified. For each technology is reported also its development status as indicated in the legend at the upper left of the image itself

## *6.2.9 Electrochemical (ECh) Glass*

Electrochemical glasses are based on photovoltaics' technology; their functioning is ensured by the electrochromic effect provided by titanium dioxide. Sunlight indeed, due to the photovoltaic effect, induces a current circulation within the material that causes dioxide's darkening. The opening of the circuit can be manually determined or by sensors, reagent to the light intensity.

### 6.2.10 Electrothermal (ET) Glass

Devices obtained through the incorporation of a heatable transparent conductive film and bus bars—to activate the film—within a glass pane. Bars are connected to the edge of the conductive film, divided into regions.

The patent of this technology has been released in 2017 to Asahi Glass Company and was related to the invention of a plate for an ET window that includes a heatable transparent conductive film and multiple bus bars for feeding electric power to the transparent conductive film. When the transparent conductive film is energized, it generates heat.

These systems are still under development; to the author's knowledge, current on-the-market-available devices are only addressed to the automotive sectors.

## 6.3 Emerging Solutions

Common approaches recently adopted to improve glazing performance, which drove the development of most of the technologies presented within the previous section (see Sects. 6.1 and 6.2), have fundamentally been four: The first can be defined as a traditional approach, aimed at acting on the interspace of a multilayer glazing assembly (IGU), by using films, low-conductance gases, or thermally improved edge spacers, to improve insulation capacity of the system. The second approach is aimed at altering material composition (such as, what happened in tinted glazing); the third approach resorts to coating application on the glass surface to modify its light reflection performance (such as selective, reflective, or Low-Emittance coatings), while the fourth approach employs external inputs of various kind (whether passive or active) to change optical features of glass.

Although valid, each of the four abovementioned approaches presents some shortcomings related to the adoption of technologies within the building sector.

To overcome existing barriers, several research centers and institutions worked or are currently working on the development of new emerging high-potential solutions. Here following are presented some of the most promising.

### 6.3.1 Bio-adaptive Glass

Bio-adaptive glazed panels are a newly emerging family of energy-generating devices, realized employing algae bio-reactor panels (Kohlstedt 2013) to produce electricity.

These bioreactors can produce biomasses, which can subsequently be collected for energy generating purposes; at the same time, they can capture solar-thermal heat, providing an energy source used to power the building. So, solar energy can be absorbed and converted or deflected, when needed, by the semi-opaque panels. The micro-algae plants contained within the glass are continuously fed with liquid nutrients and carbon dioxide through a water circuit; sunlight aids the algae photosynthesis—driving a dynamic response to the percentage of solar shading required—making them grow until they can be harvested and used as biogas to heating water.

This technology, conceived and developed from Arup in cooperation with SSC (Strategic Science Consult of Germany), has been installed for the first time in the BIQ house during the International Building Exhibition in Hamburg in 2013 (Fig. 6.23) (for further details, see Sect. 4.1.4). Despite this, to the author's knowledge, it is not currently spread in ordinary building applications.

## 6.3.2 Heat Insulation Solar Glass

HISG is a multifunctional glazing technology with a very promising U-value (Young et al. 2014; Cuce et al. 2014), resulting as a successful application of a transparent PV glazing.

The system, investigated at the University of Nottingham (Cuce et al. 2016), combines several features, such as power generation capacity, thermal insulation,

**Fig. 6.23** On the left (**a**): Arup and Slitterwerk, BIQ House—Algae Powered Building, Hamburg. Source: https://commons.wikimedia.org/wiki/File:Algae_House.JPG by Gerhard kemme (own work). CC0 (https://creativecommons.org/publicdomain/zero/1.0/deed.en) via Wikimedia Commons. On the right (**b**): Microalgae bioreactor panels' detail. Source: https://commons.wikimedia.org/wiki/File:IBA_Hamburg_BIQ_Fassadenteil_mit_Mikroalgen.nnw.jpg by NordNord West (own work). CC BY-SA 3.0 DE (https://creativecommons.org/licenses/by-sa/3.0/de/deed.en) via Wikimedia Commons

self-cleaning and acoustic insulation properties, although it is coupled with the detriment of SHG, reducing it up to 80%, and VLT (lowering it of about 7.15%), when compared to ordinary glazing.

Cuce et al. (2015) compared three typologies of conventional glazing technologies with thermal performance features of the HISG, discovering that HISG has approximately a U-value of 1.10 W/m$^2$K; this means that this technology provides better performance than conventional IGUs, whereas thermal performance parameters of air-filled double-glazed windows (Low-E treated) and Argon-filled double-glazed window (Low-E treated) are found to be similar.

### 6.3.3 Air Sandwich

Air-Sandwich glazing systems are based on the idea of a set of plastic films, with spacers and air trapped in between used as insulation. The U-value of the glazing system resulting from the assembly of five air layers is 3.40 W/m$^2$K, while a system with seven layers allows obtaining a U-Value of 1.80 W/m$^2$K (Cuce and Riffat 2015a).

### 6.3.4 Vacuum Tube Window Technology

Vacuum Tube Window technology, early presented and discussed by Eames (2008) and Schultz and Jensen (2008), was further reintroduced by Cuce and Riffat in 2015 (Cuce and Riffat 2015b), when they developed a new concept of it that provided an optimized design suitable for new applications, such as concerning existing buildings. This technology can be described as a combination of evacuated glass tubes and a glazed frame; the airgap between tubes and glass' panes is filled with Argon.

The results obtained from the experimentations show that the thermal transmittance of these systems is a direct function of tubes' diameters, with values between 0.30 and 2.00 W/m$^2$K for diameters of 80 and 20 mm, respectively.

Therefore, Vacuum Tube Windows are cost-effective (with a cost lower than 100€ per square meter) and, with high performance, they can mitigate energy consumption in buildings, providing an average U-value of 0.40 W/m$^2$K, with an entire thickness of 70 mm.

### 6.3.5 Water-Flow Window

Introduced for the first time by Chow et al. (2010), water-flow windows originated from the concept of removing the heat stored inside IGU's airgap, thanks to water flooding.

After simulations, achieved results indicate that the water flow can effectively reduce glass panes' temperature while reducing interior SGH as well as cooling energy demands.

### 6.3.6 Solar-Pond Window

Solar-pond window, developed at the University of Nottingham (Cuce and Riffat 2015a), aims to integrate into fenestration functions of lighting, heat collection, heat storage, heat preservation, and photoperiod control.

They consist of a closed window-body equipped with transparent longitudinal sections inside. In such a system, there are at least three completely isolated medium layers: the inner and the outer layers are filled with transparent thermal insulation liquid, while the interlayer is air, water, or vacuum layer.

The developed prototype has air in the interlayer and an average U-value of about 0.40 $W/m^2K$ (Cuce and Cuce 2017).

### 6.3.7 Self-Sufficient Smart Window

At the Nanyang Technological University of Singapore, Professor Xiaowei and his team developed a new smart window typology (Wang et al. 2014) able to regulate the amount of light that enters the buildings, varying its color from a transparent state to a blue state without adding energy electricity.

NTU's self-sufficient smart window can lower light penetration during the day by about 50%, going back to a transparent state during the night without energy consumption but rather producing it in surplus.

Such prototype contains a liquid electrolyte between the two glass panes, covered with Tin and Indium Oxide (ITO) commonly used in television technologies. One of the two sheets is covered with a layer of pigment called Prussian blue, while the other sheet is connected to a thin aluminum strip; these two compounds are linked with common electrical wires. When the electric circuit breaks, a chemical reaction between the Prussian blue and the electrolyte occurs, making the glass blue-colored, thus reducing the building's energy needs for air-conditioning.

### 6.3.8 Electro-Kinetic Pixel Window technology

Electro-kinetic pixel window technology, developed by researchers of the University of Cincinnati, allows to control light transmission—thanks to electro-kinetic pixels—and to fully alter the color temperature of incoming visible light.

This is possible by using two electrodes that control an electrophoretic dispersion of complementary colors (usually blue/yellow in windows' applications) characterized by opposite electrical charges. The electric field created between them causes the flow of the color particles.

The interaction of light with each electrophoretic pixel (the electrodes) depends on the position of the particles in relation to them; when there are no electrodes charged and color particles are mixed in uniform dispersion the system is dark, while where the color particles are compact toward the perimeter and the lower electrode the system is clear.

By using more than two electrodes, and by altering the applied voltages between them, different color states are possible.

Electrokinetic pixels can be easily integrated within IGU or applied to already existing windows using a roll-on coating consisting of a honeycomb of electrodes.

This patent-pending research seems able to lead to the development of low-cost window tinting, using existing e-ink technologies, that can provide a wide range of optical functionality, such as adjusting for color and brightness.

### *6.3.9 Elastomer-Deformation Tunable Window*

The elastomer-deformation tunable window is an emerging glazing technology based on mechanical mechanism able to vary their state from a clear to a tinted opaque state, without blocking the diffusion of light.

This effect is produced by the insertion of a glass pane between two transparent dielectric elastomer layers, covered with a network of electrically conducting silver nanowires. Applying a voltage, the nanowires turn into electrodes moving towards each other, thus deforming the two elastomer layers. The irregular aspect of the surface due to the chaotic distribution of the nanowires determines the refraction of light and the decreasing of the system's optical transmittance, accordingly, leaving untouched the color of the pane.

## References

Al-Obaidi KM et al. (2014) A Review of Skylight Glazing Materials in Architectural Designs for a Better Indoor Environment. Modern Applied Science 8

Allen K et al. (2017) Smart windows. Dynamic control of building energy performance. Energy and Buildings 139:535-546. https://doi.org/10.1016/j.enbuild.2016.12.093

Baetens R, Jelle BP, Gustavsen A (2011) Aerogel insulation for building applications: A state-of-the-art review. Energy and Buildings 43:761-769

Bogati S, Georg A, Graf W (2017) Photoelectrochromic devices based on sputtered WO and TiO films. Energy Materials & Solar Cells 163:170–177

Buratti C, Moretti E (2011) Transparent insulating materials for buildings energy saving: experimental results and performance evaluation. In Proceedings of the Third International Conference on Applied Energy, 16-18 May 2011, p 1421-1432

Cannavale A et al. (2015) Perovskite photovoltachromic cells for building integration. Energy & Environmental Science 8:1578-1584

Cannavale A et al. (2016) Forthcoming perspectives of photoelectrochromic devices: a critical review. Energy & Environmental Science 9:2682-2719

Carmody J et al. (2014) Window Systems for High-Performance Buildings. W.W. Norton and Company, New York

Casini M (2016) Smart Buildings. Advanced Materials and Nanotechnology to Improve Energy-Efficiency and Environmental Performance. Woodhead Publishing Series in Civil and Structural Engineering: Number 69. Elsevier, Oxford (UK)

Chow T, Li C, Lin Z (2010) Innovative solar windows for cooling-demand climate. Solar Energy Materials & Solar Cells 94: 212-220

Cuce E (2016) Toward multi-functional PV glazing technologies in low/zero carbon buildings: Heat insulation solar glass—Latest developments and future prospects. Renewable and Sustainable Energy Reviews 60:1286–1301

Cuce E, Cuce PM (2017), Solar Pond Window Technology for Energy-Efficient Retrofitting of Buildings: An Experimental and Numerical Investigation. Arab J Sci Eng 42:1901-1916

Cuce E, Cuce PM, Young EH (2016) Energy saving potential of heat insulation solar glass: Key results from laboratory and in-situ testing. Energy 97:369-380

Cuce E, Riffat SB (2015a) A state-of-the-art review on innovative glazing technologies. Renewable and Sustainable Energy Reviews 41:695-714

Cuce E, Riffat SB (2015b) Vacuum tube window technology for highly insulating building fabric: An experimental and numerical investigation. Vacuum 111:83-91

Cuce E, Young CH, Riffat SB (2014) Performance investigation of heat insulation solar glass for low-carbon buildings. Energy Conversion and Management 88:834-841

Cuce E, Young CH, Riffat SB (2015) Thermal performance investigation of heat insulation solar glass: A comparative experimental study. Energy and Buildings 86:595-600

Eames PC (2008) Vacuum glazing: current performance and future prospects. Vacuum 82:717-722

Energy Star (n.d.). Product Specification. Residential Windows, Doors and Skylights. Eligibility Criteria, Version 6.0. https://www.energystar.gov/sites/default/files/ES_Final_V6_Residential_WDS_Spec.pdf. Accessed 4 Feb 2021

Feng W et al. (2016) Gasochromic smart window: optical and thermal properties, energy simulation and feasibility analysis. Solar Energy Materials & Solar Cells 144:316–323

Giucastro FGS (2017) Smart windows of the future. The introduction of graphene in transparent photovoltaic. Paper presented at the 12th International Symposium on Renewable Energy Education - ISREE 2017, Strömstad, Sweden, 19-21 June 2017. http://www.stromstadakademi.se/sa_pdf/AAS-33.pdf Accessed 4 Feb 2021

Goia F (2012) Thermo-physical behaviour and energy performance assessment of PCM glazing system configurations: a numerical analysis. Frontiers of Architectural Research 1:341-347.

Gorgolis G, Karamanis D (2016), Solar energy materials for glazing technologies. Solar Energy Materials & Solar Cells 144:559–578

Granqvist GC (2006) Electrochromic materials: out of a niche. Nature Materials 5:89–90

Granqvist GC et al. (2009) Progress in chromogenics: new results for electrochromic and thermochromic materials and devices. Solar Energy Materials and Solar Cells 93:2032–2039

Granqvist GC et al. (2010) Advances in chromogenic materials and devices. Thin Solid Films 518:3046–3053

Hee WJ et al. (2015) The role of window glazing on daylighting and energy saving in buildings. Renewable and Sustainable Energy Reviews 42:323–343

Jelle BP et al. (2012a), Fenestration of Today and Tomorrow: A State-of-the-Art Review and Future Opportunities. Solar Energy Materials & Solar Cells 96:1–28

Jelle BP, Breivik C, Rokenes HD (2012b) Building integrated photovoltaic products: a state-of-the-art review and future research opportunities. Solar Energy Materials & Solar Cells 100:69–96.

Kim Y et al. (2015) Optical properties and optimized conditions for polymer dispersed liquid crystal containing UV curable polymer and nematic liquid crystal. Current Applied Physics 15:292–297

Kohlstedt K (2013), Algae-Fueled Building: World's First Bio-Adaptive Façade. Available via WEB Urbanist. http://weburbanist.com/2013/05/02/algae-fueled-building-worlds-first-bio-adaptive-facade/ Accessed 4 Feb 2021

Li SY, Niklasson GA, Granqvist GC (2012) Thermochromic fenestration with VO2-based materials: three challenges and how they can be met. Thin Solid Films 520:3823–3828

Liu H (2012) The development of novel window systems towards low carbon buildings. Dissertation, The University of Nottingham

Llordés A et al. (2016) Linear topology in amorphous metal oxide electrochromic networks obtained via low-temperature solution processing. Nature Materials 15:1267–1273

Long L et al. (2014) Performance demonstration and evaluation of the synergetic application of vanadium dioxide glazing and phase change material in passive buildings. Applied Energy 136:89–97

Mellot NP et al. (2006), Commercial and laboratory prepared titanium dioxide thin film for self-cleaning glasses: photocatalytic performance and chemical durability. Thin Solid Films 502:112–120

Midtdal K, Jelle BP (2013) Self-cleaning glazing products: A state-of-the-art review and future research pathways. Solar Energy Materials & Solar Cells 109:126–141

Nghana B, Tariku F (2016) Phase Change Material's impacts on the energy performance and thermal comfort of buildings in a mild climate. Build Environ 99:221–238

Peschel (2015) Transparent solar windows, Sun & Wind Energy. http://www.sunwindenergy.com/photovoltaics/transparent-solar-windows Accessed 4 Feb 2021

Qu J et al. (2014) Transparent thermal insulation coatings for energy efficient glass windows and curtain walls. Energy and Buildings 77:1–10

Rezaei SD, Shannigrahi S, Ramakrishna S (2017) A review of conventional, advanced, and smart glazing technologies and materials for improving indoor environment. Solar Energy Materials & Solar Cells 159:26–51

Schultz JM, Jensen KI (2008) Evacuated aerogel glazings. Vacuum 82:723–729

Tavares PF et al. (2014) Evaluation of electrochromic windows impact in the energy performance of buildings in Mediterranean climates. Energy Policy 67:68–81

Tucci F (2006) Involucro Ben Temperato. Efficienza energetica ed ecologica in architettura attraverso la pelle degli edifice. Alinea Editrice, Firenze

Wang J et al. (2014) A bi-functional device for self-powered electrochromic window and self-rechargeable transparent battery applications. Nature Communications 5:4921

Wheeler LM, Moore DT, Ihly R et al. (2017) Switchable photovoltaic windows enabled by reversible photothermal complex dissociation from methylammonium lead iodide. Nat Commun 8:1722. https://doi.org/10.1038/s41467-017-01842-4

Wilson HR (1995) High-Performance Windows, in McGowan A (1995) Energy-efficient residential and commercial windows reference guide

Yao J, Zhu N (2012) Evaluation of indoor thermal environmental, energy and daylighting performance of thermotropic windows. Building and Environment 49:283–290

Young CH, Chen YL, Chen PC (2014) Heat insulation solar glass ad application on energy efficiency buildings. Energy and Buildings 78:66–78

Zoller A (1924) Hollow pane of glass. German Patent 387655, 1924

# Chapter 7
# Smart Windows Technologies

## 7.1 State of the Art: A General Overview

Architectural windows that smartly regulate incident solar radiation, by changing their transmittance features in response to stimuli of various nature, are developed as a promising solution toward reducing the energy consumption of buildings.

Besides, this sector of emerging technologies, formerly appeared on the market at around the end of the last century, has received hundreds of millions of dollars in investment over the past few years; the size of the global market, which was valued at $2.8 billion in 2016, is estimated to grow to $8.35 billion in 2023, with an average increase of 16.6%.[1]

This has resulted in a rapid expansion of the production capacity globally, which, coupled with the increasing trends toward energy efficiency, comfort, and privacy, also endorsed by the most recent directives in terms of consumption and reduction in GHGs emissions, are creating demand.

Indeed, by now are clear the benefits that the adoption of smart windows technologies can provide dynamic modulation of light transmittance, promising on-demand control options, almost perfect balance between energy efficiency and aesthetics, and saving on energy costs, only to mention a few; all crucial goals for buildings' energy efficiency.

Smart window technology has undergone rapid development in the last decade: Significant efforts have been put onto windows based on electro-, thermo-, mechano-, and photoresponses (Ke et al. 2019). However, despite this, the market today remains relatively small, primarily because of the remaining challenges, such as

[1] Smart Glass Market by Technology, Application, and Geography. Available via: https://www.marketsandmarkets.com/Market-Reports/smartglass-market-907.html?gclid=Cj0KCQjwoInnBRDDARIsANBVyATYrlUxmjlbeoXxV3VtvhGrzeLZ2O5OGU4xM70bXlapuFzbxKmaI-UaAlv-SEALw_wcB. Accessed 02 Mar 2021.





V. Frighi, *Smart Architecture – A Sustainable Approach for Transparent Building Components Design*, https://doi.org/10.1007/978-3-030-77606-0_7

over price and lack of economies of scale, and confusion about knowledge of technologies, especially within the wider audience.

As a matter of fact, smart windows refer to a multitude of technologies; each of them offers a different set of features. The review performed within the previous chapter (see Chap. 6) systematically summarized the mains, including cutting-edge emerging solutions.

This further categorization, on the other hand, offers a basic assessment of each smart windows technology previously investigated, covering operating principles, advantages and disadvantages, and unresolved technology challenges, including a well-referenced benchmark of different technologies based on cost, light transmission properties, controllability, power usage if applicable, and more.

To have rigorous and reliable performance data on which base further analysis and considerations, a set of commercial products, currently market available, have been taken into consideration to obtain performance ranges that can be used as a reference, according to the requirements' framework that can vary case by case and to the different and specific application contexts.

Apparently, not all of the technologies investigated are equally widespread in the current market, especially with regard to those still under implementation (such as the emerging technologies mentioned in Chap. 6, Sect. 6.3).

This second-level analysis aims to define the main properties of actual transparent building components within the current application context (referring to real products), understanding their strength and weaknesses in the smart architecture viewpoint, and defining common terms and performance evaluation criteria to compare products with different characteristics.

To perform the abovementioned analysis, the parameters took into consideration, descended from the families of design needs and building envelope performance requirements,[2] have been defined as follows:

- *U-value*, or thermal transmittance of the system, expressed in $W/m^2K$
- *Solar Heat Gain Coefficient*, *Visual Light Transmittance* and *Light-to-Solar Gain* ratio, calculated as the ratio among the previous two parameters
- *Sound Insulation Capacity* of the system (if available), expressed in Decibel [Db] through the $R_w$ value
- *Power of the system* (in case of energy gėneration features), expressed in Watt [W]
- *Appropriateness to the Climate*
- *Operation type*
- *Degree of adaptability*
- *Trigger* (if relevant or applicable)
- *Purpose*, that led to the adoption of such a particular kind of technology

[2] Thermal control, defined as the thermal insulation capacity of the system; Visual performance, intended as the coupling of solar control ability of the system—defined as the ability to oppose the heat flux, limiting undesired gain during summer—and its light transmission control ability—intended as the ability of the system in obstruct the solar radiation within the visible spectrum—Acoustic comfort, defined as the sound insulation capacity of the system; Energy management or energy generation abilities and Interaction and user control. For further insights, see Chap. 1.

- *Adaptive function* (if relevant or applicable)
- *Operation mode*
- *Ability to get integrated with BMS*
- *Light transmission mode*

In addition to these parameters, other complementary data are reported, such as a technology's *brief description* and the use for which its adoption is suggested according to general manufactory specifications.

Concerning this second-level analysis, a synthetic comprehensive judgment of the technology under investigation has been provided, which takes into consideration the general performance features of the examined product, its applicative potentialities, and the cost–benefit ratio. Such assessment has been expressed qualitatively, thanks to three indicators that express the appropriateness of the product to specific requirements: the lower one (POOR) expresses the total absence of such requirements—or a very low value—while the highest one (GOOD) expresses the complete resolution of the critical issues identified—or a very high value.

| |
|---|
| POOR (lack of compliance) |
| FAIR |
| GOOD (complete fulfillment) |

### 7.1.1 Tinted Glass

Tinted glass commercial products are generally available in a range of colors which varies from gray/blue to gold/bronze. Blue body tinted float glasses are generally the most widespread to offer solar control while ensuring high daylight transmittance. The blue color provides solar control, thus offering aesthetic possibilities also where a low reflection is required.

The solar control properties, as well as the color density, vary depending on the pane's thickness.

Such tinted glasses can be embedded in stratified safety glass systems, heat strengthened, curved, and enameled; they can be used in a monolithic single solution of in IGUs, even coupled with a low-emissivity glass to increase thermal insulation performance of the system.

| **General performance** | |
|---|---|
| Use | Mainly service/industry or specialized |
| Climate | Generally cooling dominant climates |
| Operation type | *Not applicable* (N.A.) |

| | |
|---|---|
| Degree of adaptability | N.A. |
| Trigger (if relevant) | N.A. |
| Purpose | Thermal comfort; visual performance |
| Adaptive function (if relevant) | Prevent/reject |
| Operation mode | Passive |
| BMS | No |
| Light transmission mode | Fixed |
| **Technical data** *(6 mm + 12 mm Air + 4 mm)* | |
| U-value (W/m$^2$K) | 2.5–2.8 |
| SHGC | 39–41 |
| VLT | 47–48 |
| LSG | ~1.17–1.20 |
| **General assessment** | |
| Performance features | FAIR |
| Potentiality and applicability | FAIR |
| Integrability with BMS | POOR |
| Cost/Benefits ratio | FAIR |

### 7.1.2 Reflective Glass

Generally pyrolytic online coated or vacuum coated (offline), reflective solar control glasses are commercially available in different colors (usually clear, bronze, and gray); they offer low light transmittance and medium light reflectance, providing privacy to interior spaces while allowing the view to the outside.

Reflective glass combines a low SHGC with good reflectivity, reducing in this way the need for air conditioning.

They can be used in single glazing applications as in insulating glass units, even combined with low emissivity glass; they can also be laminated, heat strengthened, toughened, and curved.

| **General performance** | |
|---|---|
| Use | Mainly Service/Industry or Specialized |
| Climate | Generally cooling dominant climates |
| Operation type | N.A. |
| Degree of adaptability | N.A. |
| Trigger (if relevant) | N.A. |
| Purpose | Thermal comfort; visual performance |
| Adaptive function (if relevant) | Prevent/reject |
| Operation mode | Passive |
| BMS | No |

| | |
|---|---|
| Light transmission mode | Fixed |
| **Technical data** *(6 mm + 12 mm Air + 4 mm)* | |
| U-value (W/m²K) | 2.8–3.0 |
| SHGC | 45–46 |
| VLT | 30–35 |
| LSG | 0.66–0.76 |
| **General assessment** | |
| Performance features | FAIR |
| Potentiality and applicability | FAIR |
| Integrability with BMS | POOR |
| Cost/Benefits ratio | FAIR |

### *7.1.3 Selective Glasses*

High-performance glasses generally coated with a double silver layer magnetron-selective coating, applied through a cathodic pulverization vacuum, to provide thermal insulation and solar protection. Selective glass provides high-light transmission, low SHGC, and very good selectivity due to the VLT ratio always ≤2.

Commercially available in a wide range of different transmission and light-reflection level and colors, they can combine solar protection during summer and thermal insulation during winter.

| | |
|---|---|
| **General performance** | |
| Use | Service/industry; specialized |
| Climate | Any |
| Operation type | N.A. |
| Degree of adaptability | N.A. |
| Trigger (if relevant) | N.A. |
| Purpose | Thermal comfort; visual performance |
| Adaptive function (if relevant) | Prevent/reject |
| Operation mode | Passive |
| BMS | No |
| Light transmission mode | Fixed |
| **Technical data** *(6 mm + 12 mm Air + 4 mm)* | |
| U-value (W/m²K) | 2.77–3.0 |
| SHGC | 31–43 |
| VLT | 50–59 |
| LSG | 1.61–1.37 |
| **General assessment** | |
| Performance features | FAIR |

| | |
|---|---|
| Potentiality and applicability | FAIR |
| Integrability with BMS | POOR |
| Cost/Benefits ratio | GOOD |

### 7.1.4 *Low-Emissive Glass*

Offline-coated glass can reflect the long-wavelength energy generated back into the buildings. The transparent, low-emissive coating permits the transmission of the short wavelength originating from the sun.

The main features of such systems are high light transmission, neutral color, and good thermal insulation. Such products must be assembled in IGU and clearly can be combined with other glazed technologies to maximize their performance in terms of safety or acoustic insulation.

| | |
|---|---|
| **General performance** | |
| Use | Any |
| Climate | Generally heating dominant climates |
| Operation type | N.A. |
| Degree of adaptability | N.A. |
| Trigger (if relevant) | N.A. |
| Purpose | Thermal comfort |
| Adaptive function (if relevant) | Prevent/reject |
| Operation mode | Passive |
| BMS | No |
| Light transmission mode | Fixed |
| **Technical data** *(6 mm + 12 mm Air + 4 mm)* | |
| U-value ($W/m^2K$) | 1.6–1.9 |
| SHGC | 54–65 |
| VLT | 74–82 |
| LSG | 1.37–1.26 |
| **General assessment** | |
| Performance features | GOOD |
| Potentiality and applicability | FAIR |
| Integrability with BMS | POOR |
| Cost/benefits ratio | GOOD |

### 7.1.5 *Vacuum Insulated Glass (VIG)*

All purposes glass combines a very low thickness with the thermal performance of conventional double glazing. VIG (Vacuum Insulated Glazing) reaches now better performance than standard triple glazing, being suitable to be installed in almost the totality of climate regions.

They are particularly suitable for all the cases in which there is the need to balance historical preservation issues with comfort and actual environmental requirements and, in general, for all the contexts in which a thinner low-weight glazing is desirable; such systems, indeed, have the gap between the two panes reduced to 0.2 mm.

In addition, they present improved sound reduction performance when compared to a standard double-glazed unit.

| **General performance** | |
|---|---|
| Use | Any |
| Climate | Any |
| Operation type | N.A. |
| Degree of adaptability | N.A. |
| Trigger (if relevant) | N.A. |
| Purpose | Thermal comfort |
| Adaptive function (if relevant) | Control |
| Operation mode | Passive |
| BMS | No |
| Light transmission mode | Fixed |
| **Technical data** *(monolithic, 6 mm)* | |
| U-value (W/m$^2$K) | 0.7–1.4 |
| SHGC | 50–66 |
| VLT | 68–76 |
| LSG | 1.36–1.15 |
| Rw | 33–38 |
| **General assessment** | |
| Performance features | GOOD |
| Potentiality and applicability | GOOD |
| Integrability with BMS | POOR |
| Cost/Benefits ratio | FAIR |

### 7.1.6 TIM Glass

IGUs in which transparent capillary slabs are inserted in the cavity allow for a light-diffusing illumination, providing, at the same time, very good sun and glare protection. Similarly, the heat insulation capacity of the system is improved by the capillary tubes.

On the contrary, one of the weak points of Transparent Insulation Material glass is that the structure of the capillary tubes is visible from both sides, thus preventing the complete view toward outside (they are only partially transparent).

In general terms, TIM glass permits a high light transmission and good heat and sound insulation even if they partially reduce the view toward the outside.

| **General performance** | |
|---|---|
| Use | Service/industry; specialized |
| Climate | Any; heating dominant climates |
| Operation type | N.A. |
| Degree of adaptability | N.A. |
| Trigger (if relevant) | N.A. |
| Purpose | Thermal comfort |
| Adaptive function (if relevant) | Control |
| Operation mode | Passive |
| BMS | No |
| Light transmission mode | Fixed |
| **Technical data** *(6 mm + 12 mm Air + 4 mm)* | |
| U-value (W/m$^2$K) | 1.3–2.7 |
| SHGC | 26–64 |
| VLT | 22–68 |
| LSG | 0.84–1.06 |
| **General assessment** | |
| Performance features | FAIR |
| Potentiality and applicability | FAIR |
| Integrability with BMS | POOR |
| Cost/Benefits ratio | FAIR |

### 7.1.7 Self-Cleaning (SC) Glass

Float glass in which a particular, neutral coating is dropped off through a pyrolytic (online) process; the coating contains catalyst elements able to activate both the photocatalytic and the hydrophilic action. Thanks to this particular treatment, self-cleaning glasses are able to exploit solar radiation to decompose organic sediments and let them run off when rainwater droplets hit the surface, reducing cleaning and maintaining needs and related costs.

This range of glass can be laminated with both PVB and resins, heat-strengthened, curved, and enameled with standard techniques; they can be used in a monolithic form or as IGU (with the self-cleaning coating facing the outside).

| | |
|---|---|
| **General performance** | |
| Use | Any |
| Climate | Any |
| Operation type | Extrinsic |
| Degree of adaptability | On/Off |
| Trigger (if relevant) | Optical |
| Purpose | Other |
| Adaptive function (if relevant) | N.A. |
| Operation mode | Passive |
| BMS | No |
| Light transmission mode | Fixed |
| **Technical data** *(6 mm + 12 mm Air + 4 mm)* | |
| U-value (W/m$^2$K) | 1.4–2.7 |
| SHGC | 36–47 |
| VLT | 42–53 |
| LSG | 1.16–1.12 |
| **General assessment** | |
| Performance features | FAIR |
| Potentiality and applicability | FAIR |
| Integrability with BMS | POOR |
| Cost/benefits ratio | GOOD |

### *7.1.8 Heating Glass*

Commercially available heating glass generally presents a metallic surficial treatment that allows electrical stimulation of the pane; by applying an electrical current, the coating on the glass surface transforms the energy into heat, emitted in the form of long-wave infrared rays. According to coating deposition, it is possible to heat both the internal and the external pane.

The requested power supply is about 230 v–50 Hertz (to reach a surficial temperature of about 60 panes' −65 °C), and the surficial temperature is completely adjustable according to users' demands. Moreover, it is possible to add to this system, devices for the electronic management of its functioning, such as timers, thermostats, chrono-thermostats, temperature controls, automation systems, wireless systems, and so on.

The glass pane, in addition to the heating treatment, can be designed to suit other functions, such as self-cleaning functions, solar control, additional sound insulation capacity, blind insertion, and so on.

| | |
|---|---|
| **General performance** | |
| Use | Any |
| Climate | Any; heating dominant climates |
| Operation type | Users' control |
| Degree of adaptability | Gradual |
| Trigger (if relevant) | Electromagnetic |
| Purpose | Thermal comfort |
| Adaptive function (if relevant) | Prevent/reject |
| Operation mode | Active |
| BMS | Yes |
| Light transmission mode | Fixed |
| **Technical data** *(6 mm + 12 mm Air + 4 mm)* | |
| U-value ($W/m^2K$) | 1.3–2.8 |
| SHGC | 46–50 |
| VLT | 71–72 |
| LSG | 1.54–1.44 |
| W | 50–100 $m^2$ |
| **General assessment** | |
| Performance features | FAIR |
| Potentiality and applicability | GOOD |
| Integrability with BMS | GOOD |
| Cost/benefits ratio | FAIR |

### 7.1.9 *Photovoltaic (PV) Glass*

Devices that exploit a quite simple technology to capture the light of the sun and transform it into electricity.

They are made of two or more panes—generally heat-strengthened—which embed photovoltaic cells; different power capacities are determined by the solar cell density (cells' numbers per module).

The systems can be customized to adapt to each project's needs in terms of thermal insulation, solar factor, and other configurations.

| General performance | |
|---|---|
| Use | Any |
| Climate | Any |
| Operation type | N.A. |
| Degree of adaptability | N.A. |
| Trigger (if relevant) | Optical |
| Purpose | Energy generation |
| Adaptive function (if relevant) | Collect and storage |
| Operation mode | Passive (Active) |
| BMS | Yes |
| Light transmission mode | Fixed |
| **Technical data** *(6 mm + 12 mm Air + 4 mm)* | |
| U-value (W/m²K) | 1.2–2.7 |
| SHGC | 10–40 |
| VLT | 10–30 |
| LSG | 1–0.75 |
| W (Wp max per m²) | ~150–276 |
| **General assessment** | |
| Performance features | POOR |
| Potentiality and applicability | FAIR |
| Integrability with BMS | FAIR |
| Cost/Benefits ratio | FAIR |

### *7.1.10 PCM Glass*

Generally constituted by a triple insulated glass unit in which a prismatic pane is inserted in the outermost airgap to reflect back high-angle sunlight while transmitting low-angle sunlight, offering a first (passive) solar control level. The storage material—a salt hydrate PCM—is located in the inner airgap, filled with sealed polycarbonate channels into which the translucent salt is encapsulated; this provides the system a storage capacity similar to 20 cm of concrete (about 1185 Mh/m²). Such PCM melts and freezes in the temperature range of 26–30 °C.

The U-value of the system thus obtained is about 0.48 W/m²K, but its cost is still quite high (about \$550–970/m², according to the North American distribution), having a payback period of 5–10 years.

| General performance | |
|---|---|
| Use | Any |
| Climate | Any; heating dominant climates |
| Operation type | Intrinsic |

| | |
|---|---|
| Degree of adaptability | Gradual |
| Trigger (if relevant) | Thermal |
| Purpose | Thermal comfort |
| Adaptive function (if relevant) | Collect and storage; prevent/reject |
| Operation mode | Passive |
| BMS | No |
| Light transmission mode | Variable |
| **Technical data** *(6 mm + 12 mm Air + 4 mm)* | |
| U-value (W/m²K) | 0.30–0.65 |
| SHGC | 33/4–37 |
| VLT | 8–28/12–44 |
| LSG | 0.84/3–1.18 |
| **General assessment** | |
| Performance features | FAIR |
| Potentiality and applicability | FAIR |
| Integrability with BMS | POOR |
| Cost/Benefits ratio | FAIR |

### *7.1.11 SPD Glass*

The device in which the particles inserted within the thin laminated film, sandwiched between the two glass layers, react to an electric current: When no energy is applied, the particles are randomly orientated, absorbing light (so the glass appears completely opaque or colored); otherwise, when voltage is applied the particles align, reporting glass to its initial translucent state.

Their particular feature is the fact that such film is not limited to regular-shaped flat glass, thus allowing its employ with custom-shaped components.

Additionally, this SPD glass can be manually or automatically tuned to control the amount of light, glare, and heat passing through a window, reducing the need for conditioning.

| **General performance** | |
|---|---|
| Use | Any |
| Climate | Any |
| Operation type | Intrinsic; users' control |
| Degree of adaptability | Gradual |
| Trigger (if relevant) | Electromagnetic |
| Purpose | Visual performance |
| Adaptive function (if relevant) | Prevent/reject |
| Operation mode | Active |

| | |
|---|---|
| BMS | Yes |
| Light transmission mode | Variable |
| **Technical data** *(6 mm + 12 mm Air + 4 mm)* | |
| U-value (W/m$^2$K) | 1.36–1.99 |
| SHGC | 39/33 |
| VLT | 44–79/2–7 |
| LSG | 1.56/0.13 |
| Rw | 37 dB |
| **General assessment** | |
| Performance features | FAIR |
| Potentiality and applicability | FAIR |
| Integrability with BMS | GOOD |
| Cost/Benefits ratio | FAIR |

### *7.1.12 LCD Glass*

Suitable both for privacy and furniture purposes, LCD glass devices are able to frost on demand, changing from transparent to milky state. When powered indeed, panes are transparent, allowing full view and daylight, while when switched off view is obscured, providing complete privacy thus blocking about 99% of UV rays. Their voltage of operation is 110/120 VAC.

Such systems are available with various configurations and IGU thicknesses, suitable for different purposes; moreover, they are fully compatible with every kind of domotics, being controllable by switch or through a voice command.

| | |
|---|---|
| **General performance** | |
| Use | Any |
| Climate | Any |
| Operation type | Intrinsic; users' control |
| Degree of adaptability | On/Off |
| Trigger (if relevant) | Electromagnetic |
| Purpose | Other |
| Adaptive function (if relevant) | Prevent/reject |
| Operation mode | Active |
| BMS | Yes |
| Light transmission mode | Variable |
| **Technical data** *(6 mm + 12 mm Air + 4 mm)* | |
| U-value (W/m$^2$K) | ~1.59 |
| SHGC | 73/53 |
| VLT | 75/40 |

| | |
|---|---|
| LSG | 1.02/0.75 |
| Rw | 37–38 dB |
| **General assessment** | |
| Performance features | FAIR |
| Potentiality and applicability | FAIR |
| Integrability with BMS | GOOD |
| Cost/Benefits ratio | FAIR |

### *7.1.13 Electrochromic (EC) Glass*

Windows that tint in response to the variations of outdoor conditions due to the presence of nano-layers of electrochromic coating on glass' surface can respond to a low voltage (about 100–240 VAC) of 50/60 Hz.

They can adjust their tint levels based on the quantity of solar radiation allowed into the room depending on buildings' heat load specifications, personal preferences, or, again, by calculating the sun's position, thanks to light sensors and real-time data on weather conditions.

Their control system consists of a panel that communicates to windows within which the master controller controls all critical and interface functions.

| | |
|---|---|
| **General performance** | |
| Use | Any |
| Climate | Any; cooling dominant climates |
| Operation type | Extrinsic; users' control |
| Degree of adaptability | Gradual |
| Trigger (if relevant) | Electromagnetic |
| Purpose | Visual performance; thermal comfort |
| Adaptive function (if relevant) | Prevent/reject |
| Operation mode | Active |
| BMS | Yes |
| Light transmission mode | Variable |
| **Technical data** *(6 mm + 12 mm Air + 4 mm)* | |
| U-value ($W/m^2K$) | ~1.80 |
| SHGC | 40–24/4 |
| VLT | 59–33/1 |
| LSG | 1.47–1.37/0.25 |
| Rw | 35 dB |
| **General assessment** | |
| Performance features | FAIR |
| Potentiality and applicability | GOOD |
| Integrability with BMS | GOOD |
| Cost/Benefits ratio | FAIR |

### 7.1.14 *Thermochromic (TC) Glass*

These are dynamic glasses that do not need wires or controls since they work only with exposure to direct sunlight.

Indeed, thanks to the heat coming from sunlight, the thermochromic PVB layer placed within the glass panes warms, tinting accordingly in a relationship of direct proportion with the intensity of the sun. As the interlayer cools, the glass returns to a clear state.

Due to the fact that this device does not require power source or control equipment, its installation is the same as conventional glass IGUs.

| **General performance** | |
|---|---|
| Use | Any |
| Climate | Any; cooling dominant climates |
| Operation type | Extrinsic; self-adjusting |
| Degree of adaptability | Gradual |
| Trigger (if relevant) | Thermal |
| Purpose | Visual performance; thermal comfort |
| Adaptive function (if relevant) | Prevent/reject |
| Operation mode | Passive |
| BMS | No |
| Light transmission mode | Variable |
| **Technical data** *(6 mm + 12 mm Air + 4 mm)* | |
| U-value (W/m$^2$K) | ~1.36 |
| SHGC | 36/16 |
| VLT | 54/8 |
| LSG | 1.5/0.5 |
| **General assessment** | |
| Performance features | FAIR |
| Potentiality and applicability | GOOD |
| Integrability with BMS | POOR |
| Cost/Benefits ratio | FAIR |

### 7.1.15 *Photochromic (PC) Glass*

Photochromic devices are able to darken gradually when heated from direct sunlight and automatically adjust their tint level in response to the sun's intensity. Their switching process is reversible in a fully autonomous way.

However, the shortcoming of this technology is that they cannot control heat gain because the amount of incident light on the window often does not correspond to the amount of solar heat absorbed. So, for instance, the system could darken more in winter than in summer, since sun rays may strike it more intensely due to their lower height in the sky during winter months (Fig. 7.1).

| | |
|---|---|
| **General performance** | |
| Use | Any |
| Climate | Any |
| Operation type | Extrinsic; self-adjusting |
| Degree of adaptability | Gradual |
| Trigger (if relevant) | Optical |
| Purpose | Visual performance; thermal comfort |
| Adaptive function (if relevant) | Control |
| Operation mode | Passive |
| BMS | No |
| Light transmission mode | Variable |
| **Technical data** *(6 mm + 12 mm Air + 4 mm)* | |
| U-value (W/m$^2$K) | ~1.2 |
| SHGC | 10/34 |
| VLT | 6/60 |
| LSG | 0.6/1.76 |
| **General assessment** | |
| Performance features | FAIR |
| Potentiality and applicability | FAIR |
| Integrability with BMS | POOR |
| Cost/Benefits ratio | FAIR |

## 7.2 Brief Comparison and Critical Reasoning

Although the market offers different types of high-performance glazing systems and glass is ever more frequently enhanced with added functions, to increase its whole performance response as building envelope component, smart windows and, in general, emerging technologies with outside-the-box performance, are still in the early stage of development.

So far, transparent building components—even if equipped with quite high-performance—are the weakest point of building envelopes, especially in terms of thermal resistance.

Talking in performance terms about windows' essential features, it has to be said that to operate a real control on the criteria responsible for occupants' comfort, and building envelope performance as well, the most relevant parameters to be controlled are:

- the *heat transfer coefficient*, that has to be as lower as possible
- the ability in modulating the total *Solar Heat Gain Coefficient*, expressed by the proportion of direct and indirect solar radiation transmitted through the glazing

- the portion of solar visible radiation transmitted through the glazing (that is the *Visible Light Transmission*).
- The VLT has to be as high as possible even if it is associated with the increase of buildings' cooling load (Jadhav 2016); differently, the SHGC should vary according to the external climate conditions. A high SHGC is recommended for cold climate (heating demand) to maximize solar gain, while a low SHGC is suggested for hot climate (cooler demand) to avoid overheating, especially during the summer season (Kolokotsa et al. 2011).

To complete the analysis carried out within the present chapter, some recapitulatory considerations on the different technologies investigated are here presented.

- Considerable research have been invested in the development of low-energy window designs that contribute to buildings' efficiency. While significant progress has been made toward the development of cost-effective (super)insulating window designs, more limited progress has been made toward the development of cost-effective technologies for the control of solar heat gain.
- Most of the aforementioned technologies can reduce and control (passively or actively) light transmittance and SHGC.
- Double and triple IGUs constitute the majority of glazed products on the market due to their cost-effectiveness. The different types of gas inserted in the space between glass panes can significantly increase the thermal insulation performance of such products (Konroyd-Bolden and Liao 2015).
- Traditional static solar control systems are unsuitable for energy efficiency current requirements because they reduce heat loads during summer—limiting glare phenomena—but they are not able to adapt to the variability of the external conditions.
- Aerogel and multipane systems have low overall heat transfer coefficient; aerogel panels, in particular, have a high transmittance even if coupled with a reduced vision quality due to the light scattering. This high transmittance in the infrared spectrum leads to an increase in cooling loads, especially in hot climates (Ihara et al. 2015).
- Most promising and effective technologies within the domain of "static" glazing seem Self-cleaning, heating, and photovoltaic technologies, even if each of them present some shortcomings: SCs act effectively on maintenance needs, even if they are not able to operate a control on solar and visual transmittance; heating technologies seem able to provide a sort of control toward heat losses even if, similarly to previous ones, they do not operate a control on solar and light transmission, while PVs are generally limited in terms of transparency, preventing—in some cases almost completely—the vision toward the outside.
- Dynamic systems can be effectively employed to achieve significant benefits in terms of energy efficiency, especially for climatic contexts in which the temperature variates, and the weather experiences high-temperature differences. On the contrary, their high costs make them effective—in terms of cost-benefit ratio—only for great applications and/or in regions with higher energy costs, following a systemic approach (Adriaenssens et al. 2014; Xu and van Dessel 2008).

**Fig. 7.1** Summary chart of the investigated products' performance. Data and consideration reported come from the tables developed to conduct the analysis presented in the present chapter, which concerns passive and active (static and dynamic) windows technologies while regarding emerging solutions, information has been extracted from the description included in Chap. 6, Sect. 6.3. Values reported under the technical data set are referred to current widespread applications of windows technologies (6 mm + 12 Air + 4/6 mm) within the building sector

| | General performance | | | | | | | | | | | |
|---|---|---|---|---|---|---|---|---|---|---|---|---|
| | Category | Status | Use | Climate | Operation | Adaptability | Trigger | Purpose | Adaptive function | Mode | BMS | LT |
| Passive I Static | Tinted glass | Fully available | SI/Sp | T/C | not applicable | not applicable | not applicable | T/V | Prevent/Reject | Passive | No | Fixed |
| | Reflective glass | Fully available | SI/Sp | T/C | not applicable | not applicable | not applicable | T/V | Prevent/Reject | Passive | No | Fixed |
| | Selective glass | Fully available | SI/Sp | Any | not applicable | not applicable | not applicable | T/V | Prevent/Reject | Passive | No | Fixed |
| | Low-emissive glass | Fully available | Any | Tr/DD/T | not applicable | not applicable | not applicable | T | Prevent/Reject | Passive | No | Fixed |
| | Vacuum Insulated Glass | Scarcely available | Any | Any | not applicable | not applicable | not applicable | T | Control | Passive | No | Fixed |
| | TIM glass | Scarcely available | SI/Sp | T/C | not applicable | not applicable | not applicable | T | Control | Passive | No | Fixed |
| | Self-cleaning glass | Fully available | Any | Any | Extrinsic | On/Off | Optical | other | not applicable | Passive | No | Fixed |
| | Heating glass | Fully available | Any | T/C/P | Users/Control | Gradual | Electromagnetic | T | Prevent/Reject | Active | Yes | Fixed |
| | Photovoltaic glass | Fully available | Any | Any | not applicable | not applicable | Optical | E | Collect and Storage | Passive | Yes | Fixed |
| Active I Dynamic | PCM glass | Scarcely available | Any | T/C/P | Intrinsic | Gradual | Thermal | T | Collect/Prevent | Passive | No | Variable |
| | SPDs | I Fully available | Any | Any | Intrinsic/UC | Gradual | Electromagnetic | V | Prevent/Reject | Active | Yes | Variable |
| | LCDs | Fully available | Any | Any | Intrinsic/UC | On/Off | Electromagnetic | other | Prevent/Reject | Active | Yes | Variable |
| | OLEDs | Scarcely available | SI/Sp | Any | - | On/Off | Electromagnetic | other | not applicable | Active | Yes | Variable |
| | Electrochromic glass | Fully available | Any | DD/T | Extrinsic/UC | Gradual | Electromagnetic | V/T | Prevent/Reject | Active | Yes | Variable |
| | Thermochromic glass | Scarcely available | Any | DD/T | Extrinsic/SA | Gradual | Thermal | V/T | Prevent/Reject | Passive | No | Variable |
| | Thermotrophic glass | Develop./Protoype | Any | Any | Extrinsic/SA | Gradual | Thermal | V/T | Prevent/Reject | Passive | No | Variable |
| | Photochromic glass | Scarcely available | Any | Any | Extrinsic/SA | Gradual | Optical | V/T | Control | Passive | No | Variable |
| | Gasochromic glass | Develop./Protoype | - | - | Extrinsic/SA | Gradual | Other | V/T | Control | Passive | No | Variable |
| | Electrochemical glass | Develop./Protoype | - | - | Extrinsic | Gradual | Electromagnetic | V/T | Control | Passive | Yes | Variable |
| | Electrothermal glass | Develop./Protoype | - | - | Extrinsic | Gradual | Electromagnetic | V/T | Control | Passive | Yes | Variable |
| Emerging solutions | Bio-adaptive glass | Develop./Protoype | - | - | not applicable | not applicable | Optical | T/V | Collect/Prevent | Passive | No | Variable |
| | Heat Insulation Solar Glass | Research | - | - | not applicable | not applicable | not applicable | T | Prevent/Reject | Passive | No | Fixed |
| | Air Sandwich | Research | - | - | not applicable | not applicable | not applicable | T | Prevent/Reject | Passive | No | Fixed |
| | Vacuum Tube Window Tech | Research | - | - | not applicable | not applicable | not applicable | T | Prevent/Reject | Passive | No | Fixed |
| | Water-Flow Window | Research | - | - | not applicable | not applicable | not applicable | T | Prevent/Reject | Passive | No | Fixed |
| | Solar Pond Window | Research | - | - | not applicable | not applicable | not applicable | T | Prevent/Reject | Passive | No | Fixed |
| | Self-Sufficient Smart Window | Research | - | - | Extrinsic | Gradual | not applicable | V/T | Prevent/Reject | Passive | No | Variable |
| | Electro-Kinetic Pixel Window | Research | - | - | Extrinsic | Gradual | Electromagnetic | V/T | Control | Active | Yes | Variable |
| | Elastomer-Def. Tunable Window | Research | - | - | Intrinsic | Gradual | Electromagnetic | V/T | Control | Active | Yes | Variable |

| Category | Technical data* | | | | | | Assesment | | | |
|---|---|---|---|---|---|---|---|---|---|---|
| | U [W/m2K] | SHGC [%] | VLT [%] | LSG | Rw [dB] | W | 1 | 2 | 3 | 4 |
| Tinted glass | 2,5-2,8 | 39-41 | 47-48 | 1,17-1,20 | - | - | | | | |
| Reflective glass | 2,8-3,0 | 45-46 | 30-35 | 0,66-0,76 | - | - | | | | |
| Selective glass | 2,77-3,0 | 31-43 | 50-59 | 1,61-1,37 | - | - | | | | |
| Low-emissive glass | 1,6-1,9 | 54-65 | 74-82 | 1,37-1,26 | - | - | | | | |
| Vacuum Insulated Glass | 0,7-1,4 | 50-66 | 68-76 | 1,36-1,15 | 33-38 | - | | | | |
| TIM glass | 1,3-2,7 | 26-64 | 22-68 | 0,84-1,06 | - | - | | | | |
| Self-cleaning glass | 1,4-2,7 | 36-47 | 42-53 | 1,16-1,12 | - | - | | | | |
| Heating glass | 1,3-2,8 | 46-50 | 71-72 | 1,54-1,44 | - | 50-100 | | | | |
| Photovoltaic glass | 1,2-2,7 | 10-40 | 10-30 | 1-0,75 | - | 150-276 | | | | |
| PCM glass | 0,30-0,65 | 33/4-37 | 8-28/12-44 | 0,84/3-1,18 | - | - | | | | |
| SPDs | 1,36-1,99 | 39-33 | 44-79/2-7 | 1,56/0,13 | 37 | - | | | | |
| LCDs | 1,59 | 75-53 | 75-40 | 1,02/0,75 | 37-38 | - | | | | |
| OLEDs | - | - | - | - | - | - | - | - | - | - |
| Electrochromic glass | 1,8 | 40-24/4 | 59-33/1 | 1,47-1,37/0,25 | 35 | - | | | | |
| Thermochromic glass | 1,36 | 36-16 | 54-8 | 1,5-0,5 | - | - | | | | |
| Thermotrophic glass | 2,72 | 69-51 | 90-35 | 1,30-0,68 | - | - | | | | |
| Photochromic glass | 1,2 | 10-34 | 6-60 | 0,6-1,76 | - | - | | | | |
| Gasochromic glass | - | - | - | - | - | - | - | - | - | - |
| Electrochemical glass | - | - | - | - | - | - | - | - | - | - |
| Electrothermal glass | - | - | - | - | - | - | - | - | - | - |
| Bio-adaptive glass | - | - | - | - | - | - | - | - | - | - |
| Heat Insulation Solar Glass | - | - | - | - | - | - | - | - | - | - |
| Air Sandwich | - | - | - | - | - | - | - | - | - | - |
| Vacuum Tube Window Tech | - | - | - | - | - | - | - | - | - | - |
| Water-Flow Window | - | - | - | - | - | - | - | - | - | - |
| Solar Pond Window | - | - | - | - | - | - | - | - | - | - |
| Self-Sufficient Smart Window | - | - | - | - | - | - | - | - | - | - |
| Electro-Kinetic Pixel Window | - | - | - | - | - | - | - | - | - | - |
| Elastomer-Def. Tunable Window | - | - | - | - | - | - | - | - | - | - |

- means that data were not available or that consideration cannot be concluded

*Values reported are referred to current widespread application (6 mm + 12 Air + 4/6 mm)

- Electrochromic technologies give the users the ability to tune the optical transmission of the window system from clear to dark. However, they can often present a low switching speed, lack complete opacity, and offer only a mid-range price tag compared to other smart glass technologies. Thermochromic technologies, in contrast, give no control to the end user.
- Suspended Particle Devices and Liquid Cristal Devices combine high switching speed and user control. However, they fall short on controllability as only two states (clear or opaque) can be achieved. Moreover, they are almost simply used for privacy since they do not effectively block light, providing actual energy savings.
- We can say that Electrochromic, Suspended Particle Devices, and Gasochromic windows can be used in hot climates, but they all are not able to fully satisfy cool climate requirements; in addition, they still present too high costs for widespread applications: extra costs, when compared to traditional IGUs, are about 200 euros per square meter (Carmody et al. 2014), with a payback period between 26 and 33 years (DOE 2014).
- Thermochromic windows with high VLT can satisfy general requirements for hot climates even if they fail in providing requirements for temperate and cool climates (Long et al. 2015).
- Photochromic glazing has high costs; additionally, their complexity makes users unable to directly control them and obtain a uniform distribution of substances. Another weak point of such devices is the gradual loss of process' reversibility over time.
- Emerging solutions, even if equipped with very interesting potentialities under several points of view, are currently at an early stage of development, being only prototyped or tested in some small research reality; for this reason, they cannot be conceived as actually marketable solutions.

For all the abovementioned reasons, fenestrations' design optimization challenge is more complex than for many other building components and systems.

Smart glass is still often neglected in architecture and buildings, although some functions would already be technically feasible.

Research is now focusing on strategies to improve windows' performance, reducing costs and favoring sensors, and control systems installation to increase operation efficiency, making them monitorable and reactive to various stimuli based on predetermined inputs, quantifying at the same time benefits thus obtained.

Developing windows with reduced weight, cost, and thermal features, able to control and harvest energy, is of paramount importance such that they will have a high impact in the window industry.

The possibilities offered by the advanced technologies here presented, coupled with the opportunities given by the incorporation in such systems of devices based on wireless data transmission—such as, for instance, intelligent sensors capable of detecting real-time data about environmental parameters, consumption, user behavior and operation of technical plants—make possible to concretely suppose the development of a new generation of products, even smarter than before, capable of autonomous and efficient management. Additionally, obtaining feedbacks and real-time information on fenestrations' operation will lead to more informed decision makers and to the definition of more effective management strategies that will be reflected on a real implementation of the systems thus obtained (Fig. 7.2).

| | Glazing technologies | Advantages | Disadvantages |
|---|---|---|---|
| PASSIVE \| STATIC GLAZING TECHNOLOGIES | Tinted | • Low glare | • Low VLT<br>• Absorption of solar energy and release of heat into buildings. |
| | Reflective (solar control) | • Low glare<br>• Reflection of NIRs | • Low VLT |
| | Selective | • Reduction of thermal loads in summer<br>• Glare control | • Fixed coating without seasonal adaptability<br>• Modification of indoor light colour |
| | Low-Emissive | • Reduced heat re-radiation by the window | • Low SHGC |
| | Vacuum insulated | • Good thermal insulation properties<br>• Outside view<br>• Glare control | • Not uniform light distribution<br>• Ineffective in environment with temperatures' difference between internal and external greater than 35°C<br>• Poor mechanical resistance |
| | TIM | • Low thermal conductivity<br>• Glare reduction | • Lack of transparency<br>• Fragile |
| | Self-cleaning | • High-visibility capacity maintained over time | • Low control of SHGC and VLT |
| | Heating | • No condensation in the interspace<br>• Heat losses reduction<br>• Outside view | • High cost if compared to traditional "static" technologies<br>• Low control of SHGC and VLT |
| | Photovoltaic | • Solar modulation<br>• Glare control<br>• Self-powered | • Very low VLT |
| ACTIVE \| DYNAMIC GLAZING TECHNOLOGIES | PCM | • Reduced building envelope thermal fluctuations | • Lack of transparency |
| | SPDs | • Solar modulation<br>• Glare control<br>• Faster responsive time<br>• Vast transparency levels | • Need of electrical inputs for maintaining transparency levels<br>• High electric power consumption<br>• Relatively low SHGC<br>• Reduction in intensity and visibility<br>• Pretty expensive technologies |
| | LCDs | • Glare control | • Translucent (mainly for privacy applications)<br>• Need of electric power for remaining transparency<br>• Reduction in visibility<br>• Lack of SHGC modulation: only on-off state<br>• Limited to indoors<br>• High cost |
| | Electrochromics | • Solar modulation<br>• Glare control<br>• Good switching time and low power consumption | • Still expensive<br>• Limited modulation levels<br>• Long response time<br>• Need of electrical energy for very low transparent modulation<br>• Low SHGC<br>• Never completely opaque |
| | Thermochromics | • Reduced glare<br>• Reduced NIRs<br>• Reduced SHGC<br>• Relatively low price and easy to integrated into IUGs | • Low visibility<br>• High activation temperature |
| | Photochromics | • Seasonal adaptability<br>• Glare control<br>• External view<br>• UV absorption<br>• Solar and heat gain during winter | • Low absorption of IR<br>• Light-colour altering<br>• Total lack of user's control |
| | Gasochromics | • Solar modulation<br>• Glare control<br>• Faster responsive time<br>• Simple layer structure | • Not fully commercial available<br>• Need of special equipment and electrical energy for their operation |

**Fig. 7.2** Comparison between advantages and disadvantages of the most frequent aforementioned glazing technologies

# References

Adriaenssens et al. (2014) Dialectic from finding of passive and adaptive shading enclosures. Energies 7:5201-5220

Carmody J et al. (2014) Window Systems for High-Performance Buildings. W.W. Norton and Company, New York

DOE (U.S. Department of Energy) (2014), Windows and building envelope research and development: roadmap for emerging technologies. Washington, DC. https://www.energy.gov/sites/prod/files/2014/02/f8/BTO_windows_and_envelope_report_3.pdf Accessed 9 Feb 2021

Ihara T et al. (2015) Aerogel granulate glazing facades and their application potential from an energy saving perspective. Appl Energy 142:179–91

Jadhav NY (2016) Green and Smart Buildings: Advanced Technology Options. Springer, Berlin

Ke Y et al. (2019), Smart Windows: Electro-, Thermo-, Mechano-, Photochromics, and Beyond. Adv. Energy Mater. 9:1902066. https://doi.org/10.1002/aenm.201902066

Kolokotsa D et al. (2011) A roadmap towards intelligent net zero- and positive-energy buildings. Solar Energy 85:3067-3084

Konroyd-Bolden E, Liao Z (2015) Thermal window insulation, Energy and Buildings 109:245-254. https://doi.org/10.1016/j.enbuild.2015.10.005

Long L et al. (2015) Performance demonstration and simulation of thermochromic double glazing in building applications. Sol Energy 120:55–64

Xu X, Van Dessel S (2008) Evaluation of an active building envelope window system. Building and Environment 43(11):1785-91

# Chapter 8
# Transparent Building Components' Design

## 8.1 A Decision Support Tool for Early-Stage Design of Transparent Building Components

To properly compare the glazing technologies indexed in the previous chapters (see Chap. 7), a decision support tool, specifically thought for the early-stage design of Transparent Building Components, has been developed. Its final aim is to allow the setup of a sort of "open matrix"—called "Smart Windows' Configurator"—for compiling and validating different technical solutions, approaching early design problems related to transparent building components within complex building envelope systems.

The tool, which actually can be considered a toolbox, since it is composed of three different tools, started to be implemented with the systematization of data about glazing technologies gathered so far, to systematize and resume windows' technologies main features, thus, to ease the understanding of their actual applications and potentialities within the smart buildings' viewpoint.

The first item that composes it is a *balanced scorecard*, a supporting tool generally employed in management techniques to translate missions and strategies into a coherent set of actions to be implemented, connected to their measurement.

The *balanced scorecard* is born as an element of support to define the set of actions to be undertaken to increase performance features of products evaluated, concerning predetermined requirements. It connects building envelope performance requirements previously identified (the "strategic elements") with related objectives (what is expected in the domain of specific requirements) and the more operational elements that are actions to be performed on the single component (continuous improvement activities) to obtain requirements' fulfillment, according to a top-down process.



V. Frighi, *Smart Architecture – A Sustainable Approach for Transparent Building Components Design*, https://doi.org/10.1007/978-3-030-77606-0_8

Within the scorecard, KPIs that track strategic performance, as well as targets to be reached (the desired level of performance), have been expressed. For each objective, at least one KPI should be identified and tracked to indicate progress toward a desirable outcome. They will allow to measure and monitor the progress of technologies identified toward strategic targets and determine the gap between actual and targeted performance (benchmark values from standards, the basic control situation, vs. optimal strategy), thus aligning the actions to be undertaken with a shared comprehension of the objectives to be reached (Fig. 8.1).

The second tool is an *assessment matrix*, born to help in systematizing the measures depicted within the balanced scorecard with the specific objectives to perform a Multi-Criteria Decision Analysis (MCDA). It aims to support designers in analyzing complex situations by giving to each component analyzed a synthetic score that considers the initial status of it based on parameters previously identified (the KPI) according to the specific requirements. Based on such assessment, the matrix recommends which kind of actions must be undertaken to overcome the simplification of traditional components.

For each requirement, the score assigned corresponds to the *complete fulfillment* (4–5), to a *fair satisfaction* (2–3), or to the *lack of compliance* (0); an intermediate score could be possible as well (Fig. 8.2; Table 8.1).

Based on the global score achieved by each technology examined, the matrix will help in defining thresholds of intervention (operative solutions) to increase its global performance.

Three thresholds of intervention have been identified: *heavy*, which means that a hard intervention is needed; *medium*, which means that a soft intervention is needed; and *light*, which means that a low intervention is needed. A total score $\leq 10$ correspond to the heavy threshold, which means that a hard intervention is needed; a global score comprised among 11 and 19 correspond to the medium threshold, which means that soft intervention is needed while a global score higher than 20 correspond to the light threshold, meaning that a low intervention is needed.

In general terms, for each threshold of intervention, suggestions regarding performance requirements that need for implementation have been here provided (Figs. 8.3, 8.4, and 8.5):

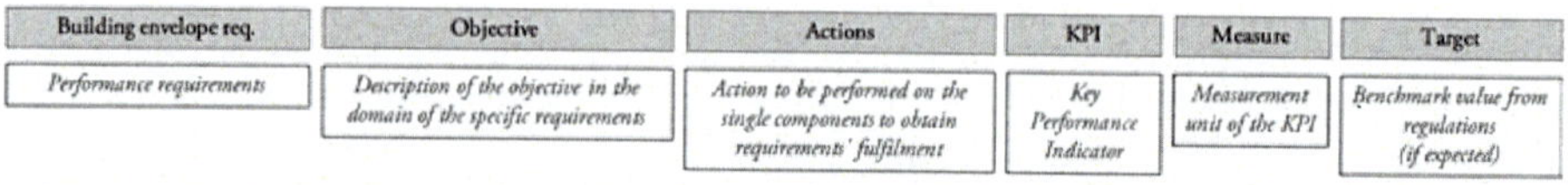


**Fig. 8.1** Structure of the balanced scorecard which defines the set of actions to be undertaken to increase performance features of technologies evaluated, concerning predetermined requirements

**Fig. 8.2** Central nucleus on which the structure of the toolbox is based

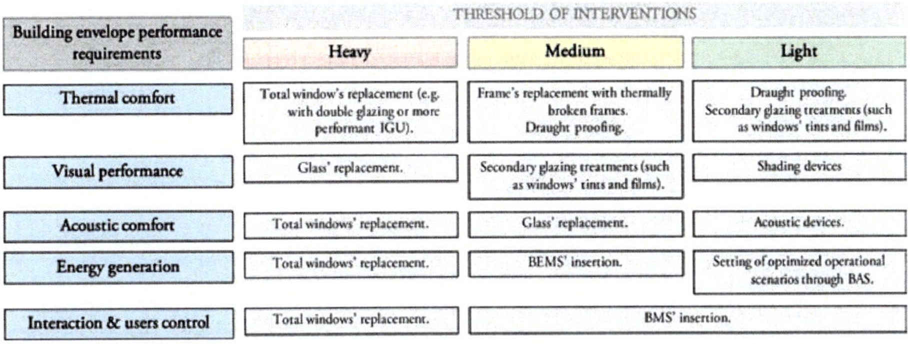


**Fig. 8.3** Suggestion regarding performance implementation needed concerning each intervention's threshold

**Table 8.1** Structure of the assessment matrix previously presented; it helps in systematizing the measures depicted within the balanced scorecard with performance objectives, giving each component a synthetic score. Based on such assessment, the matrix recommends which kind of actions must be undertaken to overcome the simplification of traditional transparent components

| Performance requirements | Parameters to be analyzed (KPI) | Glazing technology<br>*Tinted glass* |
|---|---|---|
| THERMAL COMFORT | • Thermal transmittance<br>• Appropriateness to the climate | 5 |
| VISUAL PERFORMANCE | • … | … |
| ACOUSTIC COMFORT | • … | … |
| ENERGY GENERATION | • … | … |
| … | • … | … |
| **Total score** | | **xx** |

## 8.2 Structure and Workflow

The final item of the tool(box) introduced so far is the so-called (in a non-exhaustive way) "Smart Windows' Configurator." Conceived like a sort of open matrix, it would be useful to help in compiling and validating different technical solutions, supporting the design of transparent building components toward the improvement of building envelope performance, fostering their conscious and effective integration within complex systems.

The methodology followed to conceive the "Smart Windows' Configurator" (SWC) starts with the work of cataloging of smart windows' technologies done

| Building envelope req. | Objective | Actions | KPI | Measure | Target |
|---|---|---|---|---|---|
| **Thermal comfort** | to ensure an adequate level of thermal comfort avoiding significant heat losses through transparent building components. | • number of glass' panes<br>• type of gas in the interspace<br>• installing<br>• thickness of the air gap<br>• corrective within the interspace | Thermal Transmittance [U] | W/m²K | **Basic control situation**<br>1,5 ≤ u ≤ 2<br>**Optimal strategy**<br>1 ≤ u ≤ 1,5 |
| **Visual performance** | to ensure an appropriate level of natural lightening within confined spaces avoiding undesired glare phenomena thus maintaining, where appropriate, the view towards outside. | • presence of shading devices<br>• coatings<br>• special treatments to modify optical or transmission features | Solar Control [SHGC]<br>Light Transmission [VLT] | % | **Basic control situation**<br>SHGC ≥ 40 \| VLT ≥ 70<br>**Optimal strategy**<br>SHGC ≤ 30 \| VLT ≤ 60 |
| **Acoustic comfort** | to ensure an adequate level of acoustic comfort preventing undesired noise especially in frail context of for particular intended uses. | • presence of shading devices<br>• gas in the interspace<br>• special treatments | Sound Insulation Capacity [Rw] | dB | **Basic control situation**<br>Rw = 40 dB<br>**Optimal strategy**<br>Rw ≤ 38 dB |
| **Energy generation** | (if relevant) to produce energy without compromising basic functionalities of transparent building components. | • equipment or special devices | Power generated | W | |
| **Interaction & users control** | to allow users to interact and control the device installed under several point of view | • integration with BMS | | | |
| *Performance requirements* | *Description of the objective in the domain of the specific requirements* | *Action to be performed on the single components to obtain requirements' fulfilment* | *Key Performance Indicator* | *Measurement unit of the KPI* | *Benchmark value from regulations (if expected)* |

**Fig. 8.4** Balanced scorecard filled with data coherent with the purposes of the present work; it defines the set of actions to be undertaken to increase performance features of products evaluated in relation to predetermined requirements, connecting building envelope performance requirements previously identified (the strategic elements), with related objectives (what is expected in the domain of specific requirements) and the more operational elements, which are actions to be performed on the single component (continuous improvement activities) to obtain required fulfillment, according to a top-down process

| Performance requirements | Parameters to be analized (KPI) | Passive \| Static glazing technologies | | | | | | | | |
|---|---|---|---|---|---|---|---|---|---|---|
| | | Tinted glass | Reflective glass | Selective glass | Low-E glass | VIG | TIM glass | SC glass | Heating glass | PV glass |
| THERMAL CONTROL | • Thermal transmittance<br>• Appropriateness to the climate | 3 | 3 | 3 | 5 | 5 | 5 | 3 | 5 | 3 |
| VISUAL PERFORMANCE | • SHGC<br>• VLT | 3 | 2 | 5 | 4 | 4 | 0 | 3 | 4 | 1 |
| ACOUSTIC COMFORT | • Rw | 3 | 3 | 3 | 3 | 3 | 4 | 3 | 3 | 3 |
| ENERGY GENERATION | • W | 0 | 0 | 0 | 0 | 0 | 0 | 0 | 3 | 3 |
| INTERACTION & USERS' CONTROL | • Trigger<br>• Integration with BMS | 0 | 0 | 0 | 0 | 0 | 0 | 0 | 4 | 0 |
| TOTAL SCORE | | 9 | 8 | 11 | 12 | 12 | 9 | 9 | 19 | 10 |

| Performance requirements | Parameters to be analized (KPI) | Active \| Dynamic glazing technologies | | | | | | | | | | |
|---|---|---|---|---|---|---|---|---|---|---|---|---|
| | | PCM glass | SPD glass | LCD glass | OLED glass | EC glass | TC glass | TT glass | PC glass | GC glass | ECh glass | ET glass |
| THERMAL CONTROL | • Thermal transmittance<br>• Appropriateness to the climate | 5 | 3 | 3 | - | 3 | 3 | 2 | 3 | 3 | - | - |
| VISUAL PERFORMANCE | • SHGC<br>• VLT | 1 | 4 | 2 | - | 4 | 4 | 3 | 2 | 2 | - | - |
| ACOUSTIC COMFORT | • Rw | 3 | 4 | 3 | - | 3 | 3 | 3 | 3 | 3 | - | - |
| ENERGY GENERATION | • W | 0 | 0 | 0 | - | 0 | 0 | 0 | 0 | 0 | - | - |
| INTERACTION & USERS' CONTROL | • Trigger<br>• Integration with BMS | 0 | 4 | 4 | - | 5 | 0 | 0 | 0 | 0 | - | - |
| TOTAL SCORE | | 9 | 15 | 12 | - | 15 | 10 | 8 | 8 | 8 | - | - |

| Performance requirements | Parameters to be analized (KPI) | Emerging solutions | | | | | | | | |
|---|---|---|---|---|---|---|---|---|---|---|
| | | Bio-adaptive | HISG | Air-Sandwich | VTW | WFW | SPW | SSSW | EKPW | EDTW |
| THERMAL CONTROL | • Thermal transmittance<br>• Appropriateness to the climate | 5 | 4 | 3 | 4 | 4 | 4 | 3 | 3 | 3 |
| VISUAL PERFORMANCE | • SHGC<br>• VLT | 1 | 1 | 1 | 1 | 3 | 3 | 4 | 4 | 4 |
| ACOUSTIC COMFORT | • Rw | 4 | 4 | 0 | 0 | 4 | 4 | 0 | 0 | 0 |
| ENERGY GENERATION | • W | 3 | 3 | 0 | 0 | 0 | 0 | 0 | 0 | 0 |
| INTERACTION & USERS' CONTROL | • Trigger<br>• Integration with BMS | 0 | 0 | 0 | 0 | 0 | 0 | 3 | 4 | 4 |
| TOTAL SCORE | | 13 | 12 | 4 | 5 | 11 | 11 | 10 | 11 | 11 |

*Values reported are referred to current widespread application (6mm + 12 Air + 4/6 mm)

| SCORE: | |
|---|---|
| *Complete fulfillment* | 4 - 5 |
| *Fair satisfaction* | 2 - 3 |
| *Lack of compliance* | 0 |

| THRESHOLD: | |
|---|---|
| *HEAVY* | ≤ 10 |
| *MEDIUM* | 11 - 18 |
| *LIGHT* | 19 - 24 |

**Fig. 8.5** Assessment matrix filled with data related to technologies investigated within previous chapters. Each technology has been assessed through a synthetic score that expresses its compliance with specific building envelope requirements. Based on the global score achieved by each technology examined, the matrix helps in defining thresholds of intervention to increase windows' technology global performance. Three thresholds of intervention have been identified and, for each of them, suggestions regarding performance requirements that need implementation have been provided in Fig. 8.3

within the previous chapter to understand their strengths and weaknesses to be aware of what specific implementation they need. Moreover, the analysis conducted on the Smart Architecture's sample (see Chap. 4), and more in general, on the reduced selection of those attributable to the specific field of investigation of the present work has allowed gathering data on the use of smart advanced technologies, monitoring systems, use of performance foils, or coatings able to equip Transparent Building Components (TBC) with new unconventional features, helping in fine-tune the attention toward more common critical issues when dealing with the design of TBC.

The "smart features" that emerged from such analysis have been drawn off this tool, providing useful sparks for the definition of parameters as well as for the elaboration of the outcome strategies coming from the present configurator.

The second step toward the SWC's definition was the development of the toolbox of which such configurator is part of, based on the synergy among itself and the previously developed balanced scorecard and assessment matrix (see Sect. 8.1).

In the end, the third and last step concerned the fine-tuning of the tool workflow, according to the expected results, thus the actual development of the configurator, in the form of an open matrix.

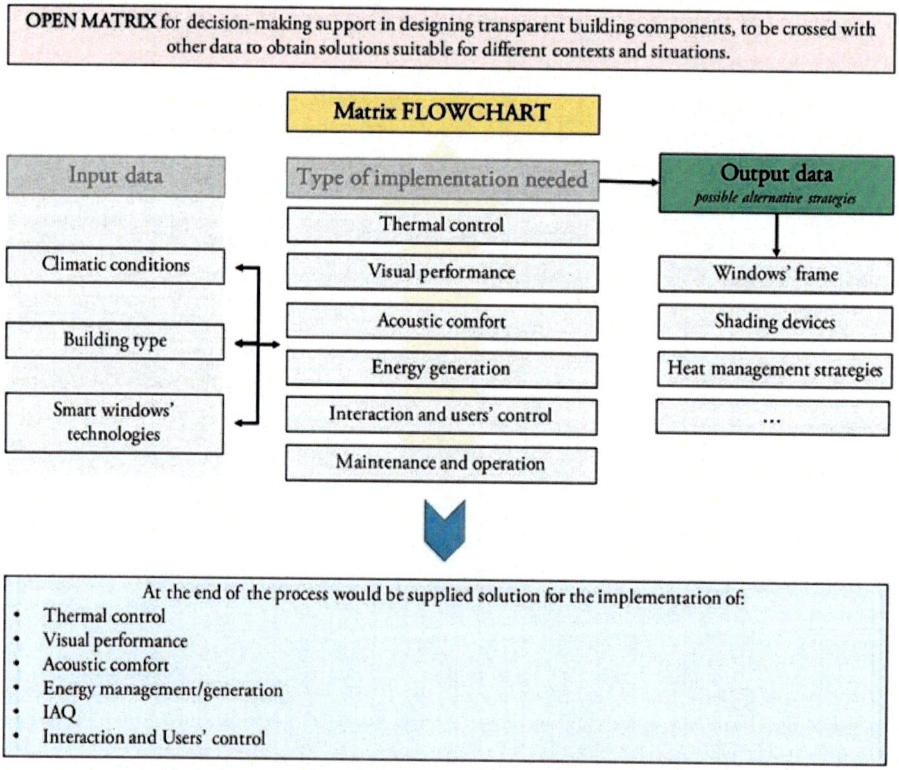


**Fig. 8.6** Flowchart for the design of the Smart Windows' Configurator. The configurator's structure is based on some input data to be assumed as a starting point (in this specific case: climatic conditions, building type, and smart windows' technologies) to be crossed with performance and technical data gathered so far about specific building envelope performance requirements to obtain as output data possible alternative strategies for each of them

Essentially, the configurator structure is based on the flowchart depicted in Fig. 8.6 which provides for some input data to be assumed as starting point; then, the matrix helps in systematizing them with performance and technical data gathered so far, concerning windows' technologies and related to building envelope performance requirements, giving—for each of them—a synthetic judgment.

Afterward, output data are provided pertinent to possible alternative strategies to solve shortcomings identified, thus supplying possible technical solutions toward the implementation of the abovementioned requirements.

Therefore, such tool aims to provide, for each scenario hypothesized, a set of viable options to be applied.

Input data on which the SWC relies on—that are reported in the vertical column of the matrix (on the y-axis)—are:

- *climate conditions*, as defined in Chap. 4, in accordance with Köppen Climate Classification, which are Tropical (Tr)/Dry-Desertic (DD)/Temperate (Tp)/ Continental (C)/Polar (P);
- *building intended use*, distinguishing among residential building (R), service/ industry building (SI), and specialized building (Sp);
- *smart windows' technologies*, referring to the technologies investigated in Chaps. 6 and 7. Within this selection, emergent solutions have been excluded due to their extremely experimental nature that made data collection very difficult about their performance characterization.

On the horizontal line instead (on the x-axis) have been depicted building envelope performance requirements, as defined in Chaps. 1 and 3, regrouped within the following "families": (a) *thermal control,* (b) *visual performance,* (c) *acoustic comfort,* (d) *energy generation,* (e) *interaction & users control*, and (f) maintenance & operation.

For each combination among the climatic condition of the surrounding, the type of buildings on which the technology has to be applied, the selected technology, and the specific requirements, the SWC "gives" a judgment able to depict synthetically the technology's behavior toward each requirement (• poor/•• fair/••• good) and, respectively, an output strategy for its implementation.

Such advice can be single, thus means referred to a single specific requirement (e.g. TrRA1, which refers to the recommendation for the implementation of thermal performance (*A*) of a tinted glazed technology (*1*) installed in a residential building (*R*) in tropical climate condition (*Tr*)) or general, thus referring to the whole implementation of the technology selected toward building envelope performance (e.g., TrR1, which refers to the recommendation for the implementation—in general terms—of a tinted glazed technology (*1*) installed in a residential building (*R*) in tropical climate condition (*Tr*)) (Fig. 8.7).

For synthesis purposes, only the combinations related to the Temperate climatic condition (Tp) and referred to Service/Industry buildings (SI) are taken into consideration because identified as the most frequent within the Smart Architecture sample analysis performed within Chap. 4.

| INPUT DATA | | TYPE OF IMPLEMENTATION NEEDED: Thermal control | Visual performance | Acoustic comfort | Energy generation | Interaction & users' control | Maintanance & operation | TOTAL |
|---|---|---|---|---|---|---|---|---|
| Climate conditions | Tr | | | | | | | |
| Building type | R | | | | | | | |
| Smart windows' technologies | | | | | | | | |
| *Tinted glass* | 1 | TrRA1 | TrRB1 | TrRC1 | TrRD1 | TrRE1 | TrRF1 | TrR1 |
| *Reflective glass* | 2 | TrRA2 | TrRB2 | TrRC2 | TrRD2 | TrRE2 | TrRF2 | TrR2 |
| *Selective glass* | 3 | TrRA3 | TrRB3 | TrRC3 | TrRD3 | TrRE3 | TrRF3 | TrR3 |
| *Low-Emissive glass* | 4 | TrRA4 | TrRB4 | TrRC4 | TrRD4 | TrRE4 | TrRF4 | TrR4 |
| *Vacuum Insulated Glass* | 5 | TrRA5 | TrRB5 | TrRC5 | TrRD5 | TrRE5 | TrRF5 | TrR5 |
| *TIM glass* | 6 | TrRA6 | TrRB6 | TrRC6 | TrRD6 | TrRE6 | TrRF6 | TrR6 |
| *Self-Cleaning glass* | 7 | TrRA7 | TrRB7 | TrRC7 | TrRD7 | TrRE7 | TrRF7 | TrR7 |
| *Heating glass* | 8 | TrRA8 | TrRB8 | TrRC8 | TrRD8 | TrRE8 | TrRF8 | TrR8 |
| *Photovoltaic glass* | 9 | TrRA9 | TrRB9 | TrRC9 | TrRD9 | TrRE9 | TrRF9 | TrR9 |
| *PCM glass* | 10 | TrRA10 | TrRB10 | TrRC10 | TrRD10 | TrRE10 | TrRF10 | TrR10 |
| *SPD glass* | 11 | TrRA11 | TrRB11 | TrRC11 | TrRD11 | TrRE11 | TrRF11 | TrR11 |
| *LCD glass* | 12 | TrRA12 | TrRB12 | TrRC12 | TrRD12 | TrRE12 | TrRF12 | TrR12 |
| *OLEDs* | 13 | TrRA13 | TrRB13 | TrRC13 | TrRD13 | TrRE13 | TrRF13 | TrR13 |
| *Electrochromic glass* | 14 | TrRA14 | TrRB14 | TrRC14 | TrRD14 | TrRE14 | TrRF14 | TrR14 |
| *Thermochromic glass* | 15 | TrRA15 | TrRB15 | TrRC15 | TrRD15 | TrRE15 | TrRF15 | TrR15 |
| *Thermotrophic glass* | 16 | TrRA16 | TrRB16 | TrRC16 | TrRD16 | TrRE16 | TrRF16 | TrR16 |
| *Photochromic glass* | 17 | TrRA17 | TrRB17 | TrRC17 | TrRD17 | TrRE17 | TrRF17 | TrR17 |
| *Gasochromic glass* | 18 | TrRA18 | TrRB18 | TrRC18 | TrRD18 | TrRE18 | TrRF18 | TrR18 |
| *Electrochemical glass* | 19 | TrRA19 | TrRB19 | TrRC19 | TrRD19 | TrRE19 | TrRF19 | TrR19 |
| *Electrothermal glass* | 20 | TrRA20 | TrRB20 | TrRC20 | TrRD20 | TrRE20 | TrRF20 | TrR20 |

**Fig. 8.7** Structure of the SWC for the first combination (Tropical climate condition—Tp, Residential building type—R). On the y-axis are reported input data on which the configurator relies on, while on the horizontal line are depicted building envelope performance requirements. For each combination, a judgment is provided—in consideration of technical data reported within the assessment matrix—able to depict in a synthetic way technology's behavior toward each requirement. Respectively, the code that refers to the output strategy for its implementation is depicted. At the end of each line instead is given the code correspondent to the suggestion related to the whole implementation of the technology selected toward building envelope performance

Moreover, particular attention is pointed toward the technologies that obtained lower score within the assessment matrix (see Sect. 8.1), for which, specifically, output strategies have been provided.

For the same reason, final outputs are only provided in general terms due to their ability to comprehend even issues related to single requirements. They combine a set of viable configurations able to inform options for optimal integration of the technical systems identified, thus becoming useful guidelines to help designers in choosing possible alternatives to overcome shortcomings of the selected technological solutions.

## 8.3 The Smart Windows' Configurator

The present section collects and presents the output strategies obtained through the SWC for the reduced sample considered; as stated before, only combinations related to the Temperate climate condition (Tp) and referred to Service/Industry buildings (SI) are taken into consideration as the most frequent within the analysis of the case studies performed in Chap. 4 (Fig. 8.8).

| INPUT DATA | | TYPE OF IMPLEMENTATION NEEDED | | | | | | | | | | | | |
|---|---|---|---|---|---|---|---|---|---|---|---|---|---|---|
| | | Thermal control | | Visual performance | | Acoustic comfrort | | Energy generation | | Interaction & users' control | | Maintanance & operation | | TOTAL |
| Climate conditions | Tp | | | | | | | | | | | | | |
| Building type | SI | | | | | | | | | | | | | |
| Smart windows' technologies | | | | | | | | | | | | | | |
| *Tinted glass* | 1 | •• | TpSIA1 | •• | TpSIB1 | •• | TpSIC1 | • | TpSID1 | • | TpSIE1 | •• | TpSIF1 | **TpSI1** |
| *Reflective glass* | 2 | •• | TPSIA2 | •• | TpSIB2 | •• | TpSIC2 | • | TpSID2 | • | TpSIE2 | •• | TpSIF2 | **TpSI2** |
| *Selective glass* | 3 | •• | TpSIA3 | ••• | TpSIB3 | •• | TpSIC3 | • | TpSID3 | • | TpSIE3 | •• | TpSIF3 | **TpSI3** |
| *Low-Emissive glass* | 4 | ••• | TpSIA4 | ••• | TpSIB4 | •• | TpSIC4 | • | TpSID4 | • | TpSIE4 | •• | TpSIF4 | **TpSI4** |
| *Vacuum Insulated Glass* | 5 | ••• | TpSIA5 | ••• | TpSIB5 | •• | TpSIC5 | • | TpSID5 | • | TpSIE5 | •• | TpSIF5 | **TpSI5** |
| *TIM glass* | 6 | ••• | TpSIA6 | • | TpSIB6 | ••• | TpSIC6 | • | TpSID6 | • | TpSIE6 | •• | TpSIF6 | **TpSI6** |
| *Self-Cleaning glass* | 7 | •• | TpSIA7 | •• | TpSIB7 | •• | TpSIC7 | • | TpSID7 | • | TpSIE7 | •• | TpSIF7 | **TpSI7** |
| *Heating glass* | 8 | ••• | TpSIA8 | ••• | TpSIB8 | •• | TpSIC8 | •• | TpSID8 | ••• | TpSIE8 | •• | TpSIF8 | **TpSI8** |
| *Photovoltaic glass* | 9 | •• | TpSIA9 | • | TpSIB9 | •• | TpSIC9 | •• | TpSID9 | • | TpSIE9 | •• | TpSIF9 | **TpSI9** |
| *PCM glass* | 10 | ••• | TpSIA10 | • | TpSIB10 | •• | TpSIC10 | • | TpSID10 | • | TpSIE10 | •• | TpSIF10 | **TpSI10** |
| *SPD glass* | 11 | •• | TpSIA11 | ••• | TpSIB11 | •• | TpSIC11 | • | TpSID11 | ••• | TpSIE11 | •• | TpSIF11 | **TpSI11** |
| *LCD glass* | 12 | •• | TpSIA12 | •• | TpSIB12 | •• | TpSIC12 | • | TpSID12 | ••• | TpSIE12 | •• | TpSIF12 | **TpSI12** |
| *Electrochromic glass* | 13 | •• | TpSIA13 | ••• | TpSIB13 | •• | TpSIC13 | • | TpSID13 | ••• | TpSIE13 | •• | TpSIF13 | **TpSI13** |
| *Thermochromic glass* | 14 | •• | TpSIA14 | ••• | TpSIB14 | •• | TpSIC14 | • | TpSID14 | • | TpSIE14 | •• | TpSIF14 | **TpSI14** |
| *Thermotrophic glass* | 15 | •• | TpSIA15 | •• | TpSIB15 | •• | TpSIC15 | • | TpSID15 | • | TpSIE15 | •• | TpSIF15 | **TpSI15** |
| *Photochromic glass* | 16 | •• | TpSIA16 | •• | TpSIB16 | •• | TpSIC16 | • | TpSID16 | • | TpSIE16 | •• | TpSIF16 | **TpSI16** |
| *Gasochromic glass* | 17 | •• | TpSIA17 | •• | TpSIB17 | •• | TpSIC17 | • | TpSID17 | • | TpSIE17 | •• | TpSIF17 | **TpSI17** |

**Fig. 8.8** Section of the Smart Windows' Configurator related to the combination for Service/ Industry buildings (SI) in Temperate climate condition (Tp). For each technology examined (emerging technologies, elechtro-chemical and elechtro-thermal technologies have been excluded due to their extremely experimental nature, thus for the impossibility to gather data on their performance and behaviors to elaborate further critical considerations.), a judgment able to depict in a synthetic way technology's behavior toward each requirement is provided based on the general description of technologies investigated, analyzed qualitatively, and enriched by knowledge acquired, thanks to performance range analysis. For each combination, the configurator gives an output strategy for the implementation of the specific window technology investigated, supporting the design of transparent building components thus fostering their integration within complex systems. Such suggestion can be single, thus mean referred to a single specific requirement (e.g., TrRA1) or general, thus means referred to the whole implementation of the technology selected toward building envelope performance (e.g., TrR1)

| Performance requirements | Parameters to be analized (KPI) | Passive \| Static glazing technologies | | | | | | | | | Active \| Dynamic glazing technologies | | | | | | | | | | |
|---|---|---|---|---|---|---|---|---|---|---|---|---|---|---|---|---|---|---|---|---|---|
| | | Tinted glass | Reflective glass | Selective glass | Low-E glass | VIG | TIM glass | SC glass | Heating glass | PV glass | PCM glass | SPD glass | LCD glass | OLED glass | EC glass | TC glass | TT glass | PC glass | GC glass | ECh glass | ET glass |
| THERMAL CONTROL | • Thermal transmittance<br>• Appropriateness to the climate | 3 | 3 | 3 | 5 | 5 | 5 | 3 | 5 | 3 | 5 | 3 | 3 | - | 3 | 3 | 2 | 3 | 3 | - | - |
| VISUAL PERFORMANCE | • SHGC<br>• VLT | 3 | 2 | 5 | 4 | 4 | 0 | 3 | 4 | 1 | 1 | 4 | 2 | - | 4 | 4 | 3 | 2 | 2 | - | - |
| ACOUSTIC COMFORT | • Rw | 3 | 3 | 3 | 3 | 3 | 4 | 3 | 3 | 3 | 3 | 4 | 3 | - | 3 | 3 | 3 | 3 | 3 | - | - |
| ENERGY GENERATION | • W | 0 | 0 | 0 | 0 | 0 | 0 | 0 | 3 | 3 | 0 | 0 | 0 | - | 0 | 0 | 0 | 0 | 0 | - | - |
| INTERACTION & USERS' CONTROL | • Trigger<br>• Integration with BMS | 0 | 0 | 0 | 0 | 0 | 0 | 0 | 4 | 0 | 0 | 4 | 4 | - | 5 | 0 | 0 | 0 | 0 | - | - |
| TOTAL SCORE | | 9 | 8 | 11 | 12 | 12 | 9 | 9 | 19 | 10 | 9 | 15 | 12 | - | 15 | 10 | 8 | 8 | 8 | - | - |

**Fig. 8.9** Shred of the assessment matrix with enlightened the windows' technologies for which output strategies are provided; the selected technologies are those that obtained lower score (lower than 10), thus for which a heavy intervention is needed

| INPUT DATA | | TYPE OF IMPLEMENTATION NEEDED | | | | | | | | | | | | |
|---|---|---|---|---|---|---|---|---|---|---|---|---|---|---|
| | | Thermal control | | Visual performance | | Acoustic comfrort | | Energy generation | | Interaction & users' control | | Maintanance & operation | | TOTAL |
| Climate conditions | Tp | | | | | | | | | | | | | |
| Building type | SI | | | | | | | | | | | | | |
| Smart windows' technologies | | | | | | | | | | | | | | |
| *Tinted glass* | 1 | •• | TpSIA1 | •• | TpSIB1 | •• | TpSIC1 | • | TpSID1 | • | TpSIE1 | •• | TpSIF1 | TpSI1 |
| *Reflective glass* | 2 | •• | TPSIA2 | •• | TpSIB2 | •• | TpSIC2 | • | TpSID2 | • | TpSIE2 | •• | TpSIF2 | TpSI2 |
| *Selective glass* | 3 | •• | TpSIA3 | ••• | TpSIB3 | •• | TpSIC3 | • | TpSID3 | • | TpSIE3 | •• | TpSIF3 | TpSI3 |
| *Low-Emissive glass* | 4 | ••• | TpSIA4 | ••• | TpSIB4 | •• | TpSIC4 | • | TpSID4 | • | TpSIE4 | •• | TpSIF4 | TpSI4 |
| *Vacuum Insulated Glass* | 5 | ••• | TpSIA5 | ••• | TpSIB5 | •• | TpSIC5 | • | TpSID5 | • | TpSIE5 | •• | TpSIF5 | TpSI5 |
| *TIM glass* | 6 | ••• | TpSIA6 | • | TpSIB6 | ••• | TpSIC6 | • | TpSID6 | • | TpSIE6 | •• | TpSIF6 | TpSI6 |
| *Self-Cleaning glass* | 7 | •• | TpSIA7 | •• | TpSIB7 | •• | TpSIC7 | • | TpSID7 | • | TpSIE7 | •• | TpSIF7 | TpSI7 |
| *Heating glass* | 8 | ••• | TpSIA8 | ••• | TpSIB8 | •• | TpSIC8 | •• | TpSID8 | ••• | TpSIE8 | •• | TpSIF8 | TpSI8 |
| *Photovoltaic glass* | 9 | •• | TpSIA9 | • | TpSIB9 | •• | TpSIC9 | •• | TpSID9 | • | TpSIE9 | •• | TpSIF9 | TpSI9 |
| *PCM glass* | 10 | ••• | TpSIA10 | • | TpSIB10 | •• | TpSIC10 | • | TpSID10 | • | TpSIE10 | •• | TpSIF10 | TpSI10 |
| *SPD glass* | 11 | •• | TpSIA11 | ••• | TpSIB11 | •• | TpSIC11 | • | TpSID11 | ••• | TpSIE11 | •• | TpSIF11 | TpSI11 |
| *LCD glass* | 12 | •• | TpSIA12 | •• | TpSIB12 | •• | TpSIC12 | • | TpSID12 | ••• | TpSIE12 | •• | TpSIF12 | TpSI12 |
| *Electrochromic glass* | 13 | •• | TpSIA13 | ••• | TpSIB13 | •• | TpSIC13 | • | TpSID13 | ••• | TpSIE13 | •• | TpSIF13 | TpSI13 |
| *Thermochromic glass* | 14 | •• | TpSIA14 | ••• | TpSIB14 | •• | TpSIC14 | • | TpSID14 | • | TpSIE14 | •• | TpSIF14 | TpSI14 |
| *Thermotrophic glass* | 15 | •• | TpSIA15 | •• | TpSIB15 | •• | TpSIC15 | • | TpSID15 | • | TpSIE15 | •• | TpSIF15 | TpSI15 |
| *Photochromic glass* | 16 | •• | TpSIA16 | •• | TpSIB16 | •• | TpSIC16 | • | TpSID16 | • | TpSIE16 | •• | TpSIF16 | TpSI16 |
| *Gasochromic glass* | 17 | •• | TpSIA17 | •• | TpSIB17 | •• | TpSIC17 | • | TpSID17 | • | TpSIE17 | •• | TpSIF17 | TpSI17 |

**Fig. 8.10** Shred of the Smart Windows' Configurator enlightened the configurations for which output strategies are provided

For synthesis purposes, within this framework, final considerations are provided in general terms only for technologies that obtained the lower score within the assessment matrix (Fig. 8.5), which means lower than 10; that is, specifically, tinted, reflective, TIM, Self-Cleaning, PCM, Thermotrophic, Photochromic, and Gasochromic (Figs. 8.9 and 8.10).

## 8.3.1 Output Strategies

Before providing general output strategies for each technology identified, some brief considerations about the specific combination under examination are needed.

First of all, it is worth remembering that the performance of glass in temperate climate must ensure solar control, thus reducing overheating during summer, allowing a high visual transmittance, and maintaining—where possible—benefits of passive solar heating. Then, VLT and energy transmission can be higher than those requested in hot climate conditions.

To allow passive solar strategies, the performance range for each value should be:

- VLT between 35 and 90% (and ideally around 70%),
- SHGC between 20 and 70% (ideally around 50%),
- U-value between 1 and 2 W/m$^2$K (anyway lower than 2.5 W/m$^2$K) with respect to specific local regulations on the subject.
- A high LSG ratio is considered as an ideal window indicator because, while controlling transmitted infrared solar radiation, the window should be able to also transmit as much visible light as possible to reduce lighting loads (Correa Almanza 2004).

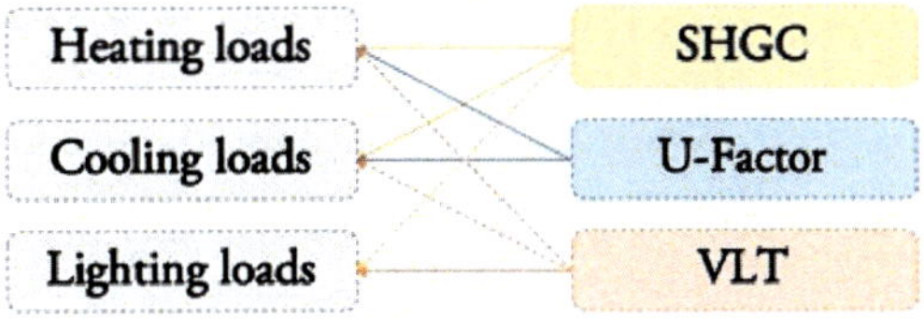


**Fig. 8.11** Correlation of the three abovementioned values and building loads. Author's elaboration on Wang et al. (2016)

Each of these recommendations must be evaluated with the specific application contexts (Fig. 8.11).

| | |
|---|---|
| **TpSI1** | **<u>Tinted glass</u>**<br>*Tinted glazing has generally different coloration (gray, green, bronzed, or blue are the most common) obtained directly within the vitreous paste; its solar control properties and its chromatic tone varies according to pane's thickness while its reflective ability is slightly low than those of clear float glass.*<br>*Thermal control* \| Quite low if relying only on the ability of the single glass pane; the tinted mass can block lighter than heat, meaning they can reduce only minimal cooling load (significant in temperate climate conditions).<br>Instead, if employed in IGUs, their insulation ability increases. In these solutions, the tinted glass should face the outside so that it can dissipate the absorbed radiation more easily. Furthermore, adding to glass panes low emissivity properties (such as, for instance, through a low-emissivity coating or film) could help in reducing energy consumption for cooling.<br>*Visual performance* \| Tinted glass reduces glare, lowering at the same time VLT, this means they are particularly suitable for highly glare-sensitive conditions even if they can diminish views and create gloomy interiors under some weather conditions. For this reason, their adoption in SI buildings must be carefully evaluated according to the activities conducted within confined spaces.<br>A good approach for daylighting and glare control could be separate view and light windows (i.e., use high-transmission clear glazing from one side, and lower transmission glazing windows, to control glare, from the other).<br>*Acoustic comfort* \| The acoustic comfort ensured by the single pane is fair and comparable to those of a common clear single glass; to obtain significant improvement, IGU systems or assembly of more than one panes with PVB could help in increasing their acoustic performance as well as recurring to particular sealing and frame technologies (with increased insulation properties) or, again, coupling them with specific acoustic devices.<br>*Energy generation* \| They do not possess energy generation ability. However, it must be observed that the energy savings coming from their adoption is greater than the energy needed for their production.<br>*Interaction and users' control* \| Tinted glass does not allow forms of interaction and users' control as they are passive, static window technologies. Limited interactions can be obtained by their integration with other devices (such as movable shading devices), suitable to be adjusted according to users' preferences or particular behavior scenarios.<br>*Maintenance and operation* \| Advantages in adopting such technology are related to their great availability and affordability.<br>In general terms, it can be said that to improve the specific technologies' performance, designers can alternate system selection opportunities with various glazing and shading options; fine-tune windows' sizing, windows' location, and shading strategy for a more efficient system; curate the installing phase to ensure the performance maintenance within the interface according to the specific wall support as well the windows' technical features (frames, fastening systems, added mechanical equipment, and other devices). |

| | |
|---|---|
| **TpSI2** | <u>**Reflective glass**</u><br>*Reflective glass presents the ability to reflect infrared radiations, reducing transmission through them without shading the infrared waves; its reflectivity increases according to the thickness of the surficial treatment (film or coating).*<br>*Thermal control* I Quite low if relying only on the ability of the single glass pane. If employed in IGUs their insulation ability increase. In these solutions, the reflective glass should face the outside to perform to its ability to reflect infrared radiation.<br>*Visual performance* I They allow only a one-side view, lowering light transmittance. The more reflective they are, the low light transmission they allow, bringing with it all the interior disadvantages that may be associated with that characteristic. However, their high lighting reflection values in the visible range (RL about 35/40%) and low in the infrared range, combined with their SHGC, recommend their use in climate contexts in which solar gains' reduction is fundamental (such as temperate climate). Moreover, it must be observed, speaking about SI buildings, that their application, as one-side view systems, allows privacy of the indoor spaces.<br>*Acoustic comfort* I The acoustic comfort ensured by the single pane is fair and comparable to those of a common clear single glass; to obtain significant improvement, IGU systems or assembly of more than one pane with PVB could help in increasing their acoustic performance as well as recurring to particular sealing and frame technologies (with increased insulation properties) or, again, coupling them with specific acoustic devices.<br>*Energy generation* I They do not possess energy generation ability.<br>*Interaction and users' control* I Reflective glass does not allow forms of interaction and users' control as they are passive, static window technologies. Limited interactions can be obtained by their integration with other devices (such as movable shading devices), suitable to be adjusted according to users' preferences or particular behavior scenarios.<br>*Maintenance and operation* I Advantages in adopting such technology are related to their great availability and affordability.<br>In general terms it can be said that to improve the specific technologies' performance, designers can alternate system selection opportunities with various glazing and shading options; fine-tune windows' sizing, windows' location, and shading strategy for a more efficient system; curate the installation phase to ensure the performance maintenance within the interface according to the specific wall support as well the windows' technical features (frames, fastening systems, added mechanical equipment, and other devices). |

| | |
|---|---|
| **TpSI6** | **TIM glass**<br>*Systems that incorporate Transparent or Translucent Insulating Material; their main advantage is that they combine diffused lighting features with limited thermal losses that usually characterize glazing surfaces.*<br>*Thermal control* \| Good heat-insulation capacity thanks to the presence of TIM in the interspace; moreover, the good protection from the sun increases their thermal performance.<br>On the contrary, the high transmittance in the infrared spectrum could lead to an increase in cooling loads during summer; for this reason, their integration with shading devices could be useful.<br>*Visual performance* \| They present high and uniform light transmittance into confined spaces, even if coupled with a reduced vision quality due to the light scattering; being translucent in fact (SHGC is equal to that of a common IGU), they allow the diffusion of natural daylight, preventing glare phenomena. However, they avoid complete transparency toward the outside.<br>*Acoustic comfort* \| Their acoustic abilities are pretty good due to the presence of the TIM within the airgap; it has to be noticed that they imply their arrangement in an IGU, thus means that their Sound Insulation Capacity does not decrease under predetermined values. Acoustic improvement could be obtained recurring to particular sealing and frame technologies (with increased insulation properties) or coupling them with specific acoustic devices.<br>*Energy generation* \| They do not possess energy generation ability.<br>*Interaction and users' control* \| TIM glasses do not allow forms of interaction and users' control as they are passive, static window technologies. Limited interactions can be obtained by their integration with other devices, suitable to be adjusted according to users' preferences or particular behavior scenarios.<br>*Maintenance and operation* \| They do not require particular precaution, even if their duration is connected to the performance maintenance of the Transparent Insulation Material employed within the airgap. However, it has to be highlighted the fact that they present a quite high cost when compared to current glazing technologies.<br>In general terms, it can be said that to improve the specific technologies' performance, designers can fine-tune windows' sizing, windows' location, and shading strategy for a more efficient system; adopt frames and devices with improved performance, curate the installing phase to ensure the performance maintenance within the interface according to the specific wall support as well the windows' technical features (frames, fastening systems, added mechanical equipment, and other devices). |

| | |
|---|---|
| **TpSI7** | **<u>Self-Cleaning glass</u>**<br>*Windows' technologies able to reduce maintenance needs, particularly suitable for situations where glazed surfaces are difficult to reach.*<br>*Thermal control* I Quite low if relying only on the ability of the single glass pane. If employed in IGUs, their insulation ability increases. Furthermore, adding to glass panes low emissivity properties (such as, for instance, through a low-emissivity coating or film) could help in reducing energy consumption for cooling.<br>*Visual performance* I Thanks to the reduction in dirt on their surfaces, self-cleaning systems improve light transmission through them, reducing energy costs for lighting accordingly. However, it has to be said that they effectively act on maintenance needs, but they are not able to operate a control on solar and visual transmittance for which shading devices or secondary glazing treatments are needed.<br>*Acoustic comfort* I The acoustic comfort ensured by the single pane is fair and comparable to those of a common clear single glass to obtain significant improvement, IGU systems, or assembly of more than one pane with PVB could help in increasing their acoustic performance as well as recurring to particular sealing and frame technologies (with increased insulation properties) or coupling them with specific acoustic devices.<br>*Energy generation* I They do not possess energy generation ability.<br>*Interaction and users' control* I SC glass does not allow forms of interaction and user control as they are passive, static window technologies. Limited interactions can be obtained by their integration with other devices (such as shading devices), suitable to be adjusted according to users' preferences or particular behavior scenarios.<br>*Maintenance and operation* I They do not require particular precaution even if their duration is limited over time; moreover, it has to be said that their effectiveness is connected with the possibility to activate the phenomenon that permits their functioning. The reason to adopt them is generally connected with the need to reduce maintenance needs of out-of-reach glazed surfaces.<br>In general terms, it can be said that to improve the specific technologies' performance, designers can alternate system selection opportunities with various glazing and shading options; fine-tune windows' sizing, windows' location, and shading strategy for a more efficient system; consider the possibility to couple this technology with secondary glazing treatments (such as windows' tints and films); curate the installing phase to ensure the performance maintenance within the interface according to the specific wall support as well the windows' technical features (frames, fastening systems, added mechanical equipment, and other devices). |

| | |
|---|---|
| **TpSI10** | **PCM glass**<br>*Glazing systems that integrate the insulating and the overheating control function in a single functional unit, thanks to the presence of a PCM mixture within the airgap of an IGU.*<br>*Thermal control* I Good heat-insulation capacity, thanks to the presence of PCMs in the interspace, which also increase passive solar heat gain. Advantages of such systems are that, in addition to the contribution given by the PCM, they generally are employed in multiple glass units, increasing evermore their thermal control ability. Furthermore, to improve passive solar control, prismatic pane or shading systems can be integrated within one of the airgap of the system.<br>*Visual performance* I They allow diffused natural lighting of spaces (about 28% at the solid state while over 40% in the liquid state), exploiting natural daylighting even during the summer season, without excessive overheating. However, it has to be said that although translucent PCMs are available, the light quality transmittance of these technologies is still a weak point.<br>*Acoustic comfort* I Their acoustic abilities are pretty good due to the presence of the PCM within the airgap; it has to be noticed that this implies their arrangement in an IGU, thus means that their Sound Insulation Capacity does not decrease under predetermined values. Acoustic improvement could be obtained recurring to particular sealing and frame technologies (with increased insulation properties) or coupling them with specific acoustic devices.<br>*Energy generation* I They do not possess energy generation abilities; however, it has to be said that the presence of the PCM confers the system the storage ability.<br>*Interaction and users' control* I PCM glasses do not allow forms of interaction and users' control although they are considered as active, dynamic window technologies. Their operation mode indeed is completely intrinsic, being only activated through external temperature variation. Limited forms of interaction can be obtained by their integration with other devices, suitable to be adjusted according to users' preferences or particular behavior's scenarios.<br>*Maintenance and operation* I They do not require particular precaution even if their duration is connected to the performance maintenance of the Phase-Change Material employed within the airgap. It has to be highlighted the fact that they present a quite high cost when compared to current glazing technologies as well as those limitations are existing regarding their large-scale applications.<br>In general terms, it can be said that to improve the specific technologies' performance, designers can fine-tune windows' sizing and windows' location; adopt frames and devices with improved performance, curate the installing phase to ensure the performance maintenance within the interface according to the specific wall support as well the windows' technical features (frames, fastening systems, added mechanical equipment, and other devices). |

| | |
|---|---|
| **TpSI15** | **Thermotropic glass**<br>*Thermotropic glass dynamically alters its solar and visible light transmittance and reflectance according to windows' temperature changes by switching between a transparent and an opaque state.*<br>*Thermal control* I Quite low if relying only on the ability of the single glass pane, even if they are applied as IGUs increase their insulation abilities. Their particular functioning protects against overheating and for use as a blind; indeed, the dynamic adjustment to the fluctuating amount of light and heat, allowing the efficient use of daylight during winter and preventing excessive heat during summer, thus reducing heating and cooling costs.<br>Dynamic systems can be effectively employed to achieve significant benefits in terms of energy efficiency, especially for climate contexts in which the temperature experiences high differences such as temperate climate.<br>*Visual performance* I By switching between a transparent and an opaque state, they offer self-regulating solar protection.<br>Even if the degree of light scattering can be varied, they do not retain visibility in the opaque mode, so as they are suitable for all application in which a direct view is not significant.<br>*Acoustic comfort* I The acoustic comfort ensured by the system is fair and comparable to those of a common IGU, which means that their Sound Insulation Capacity does not decrease under predetermined values. Acoustic improvement could be obtained recurring to particular sealing and frame technologies (with increased insulation properties) or coupling them with specific acoustic devices.<br>*Energy generation* I They do not possess energy generation ability.<br>*Interaction and users' control* I They do not allow forms of interaction and users' control although they are considered as active, dynamic window technologies. Their operation mode indeed is completely intrinsic, being only activated through external temperature variation; indeed, they do not require an electrical power source nor electrical control device are needed.<br>Limited forms of interaction can be obtained by their integration with other devices suitable to be adjusted according to users' preferences or particular behavior scenarios. Combination with an individual switching behavior is possible.<br>*Maintenance and operation* I They do not require particular precautions as they are very durable and compatible with existing conventional glazing systems, since no sealing frames are required.<br>It has to be highlighted the fact that they present a quite high cost with respect to current glazing technologies; for this reason, their application is suggested only for particular great applications or in regions with high energy costs.<br>In general terms, it can be said that to improve the specific technologies' performance, designers can fine-tune windows' sizing and windows' location; adopt frames and devices with improved performance, curate the installing phase to ensure the performance maintenance within the interface according to the specific wall support as well the windows' technical features (frames, fastening systems, added mechanical equipment, and other devices). |

| | |
|---|---|
| **TpSI16** | **Photochromic glass**<br>*Glazing systems that gradually vary their optical properties due to external light intensity variations; modify their transmission capacity based on the intensity, duration, and nature of the incident solar radiation.*<br>*Thermal control* ǀ Quite low if relying only on the ability of the single glass pane, even when they are applied as IGUs increase their insulation abilities. Their particular functioning protects against overheating and for use as a blind; indeed, the dynamic adjustment to the fluctuating amount of light and heat allows the efficient use of daylight during winter and prevents excessive heat during summer, thus reducing heating and cooling costs.<br>Dynamic systems can be effectively employed to achieve significant benefits in terms of energy efficiency, especially for climate contexts in which the temperature experiences high differences, such as temperate climate.<br>*Visual performance* ǀ Their light transmission ability depends on the amount of light on the glass surface; however, since they darken gradually when heated from direct sunlight, they cannot control heat gain (e.g., they could darken more in winter than in summer). In general, when they decrease light transmission, even SHG in the internal environment decreases.<br>*Acoustic comfort* ǀ The acoustic comfort ensured by the system is fair and comparable to those of a common IGU, which means that their Sound Insulation Capacity does not decrease under predetermined values. Acoustic improvement could be obtained recurring to particular sealing and frame technologies (with increased insulation properties) or coupling them with specific acoustic devices.<br>*Energy generation* ǀ They do not possess energy generation ability.<br>*Interaction and users' control* ǀ They do not allow forms of interaction and users' control although they are considered as active, dynamic window technologies as their operation mode are completely intrinsic, being only activated through external temperature variation. Limited forms of interaction can be obtained by their integration with other devices, suitable to be adjusted according to users' preferences or particular behavior's scenarios.<br>*Maintenance and operation* ǀ They do not require particular precautions even if their use in buildings is still poor due to the fact that they are ambient temperature sensitive. Additionally, it has to be highlighted the fact that they present a quite high cost when compared to current glazing technologies; for this reason, their application is suggested only for particular applications.<br>In general terms, it can be said that to improve the specific technologies' performance, designers can fine-tune windows' sizing and windows' location; adopt frames and devices with improved performance, curate the installing phase to ensure the performance maintenance within the interface according to the specific wall support as well the windows' technical features (frames, fastening systems, added mechanical equipment, and other devices). |

| | |
|---|---|
| **TpSI17** | **<u>Gasochromic glass</u>**<br>*Systems obtained through the incorporation of a gasochromic layer between the two panes of an IGU, which, when exposed to diluted hydrogen gas, colors, and change system's transparency according to the volume of hydrogen to which the layer has been exposed.*<br>*Thermal control* I Quite low if relying only on the ability of the single glass pane, even if they're applied as IGUs (generally double or triple), often coupled with low-emissive coatings, significantly increase their insulation abilities. Their particular functioning protects against overheating and for use as a blind; indeed, the dynamic adjustment to the fluctuating amount of light and heat allows the efficient use of daylight during winter and prevents excessive heat during summer, thus reducing heating and cooling costs.<br>Dynamic systems can be effectively employed to achieve significant benefits in terms of energy efficiency especially for climate contexts in which the temperature experiences high differences such as temperate climate.<br>*Visual performance* I Their light transmission ability depends on the volume of hydrogen to which the layer has been exposed. Contrary to other dynamic technologies, gasochromics retain clear visibility from inside to outside in all the switching states. Lower SHG values can be obtained by adding solar control coatings.<br>*Acoustic comfort* I The acoustic comfort ensured by the system is fair and comparable to those of a common IGU, which means that their Sound Insulation Capacity does not decrease under predetermined values. Acoustic improvement could be obtained recurring to particular sealing and frame technologies (with increased insulation properties) or coupling them with specific acoustic devices.<br>*Energy generation* I They do not possess energy generation ability.<br>*Interaction and users' control* I The control unit requested by the system permits both manual as automatic control; furthermore, the integration into a bus system allows the glazing to be switched to optimize lighting conditions, thermal comfort, and building energy consumption.<br>*Maintenance and operation* I They do not require particular precautions even if their use in buildings is still poor since they are pretty expensive and that they need quite complex control equipment.<br>In general terms, it can be said that to improve the specific technologies' performance, designers can fine-tune windows' sizing and windows' location; adopt frames and devices with improved performance, curate the installing phase to ensure the performance maintenance within the interface according to the specific wall support as well the windows' technical features (frames, fastening systems, added mechanical equipment and other devices). Additionally, installing daylight-activated (automated) controls can have cost-saving applications and benefits by dimming or turning off windows with them. |

# References

Correa G, Almanza R (2004), Copper based thin films to improve glazing for energy-savings in buildings. Solar Energy, 76(1):111-115

Wang JJ et al. (2016) Simulation-Based Optimization of Window Properties Based on Existing Products. Paper presented at the Building Simulation & Optimisation (BSO) Conference, Newcastle, September 2016

# Chapter 9
# A Sustainable Approach for Design of Transparent Building Components: Conclusions and Future Perspectives

The present work was motivated by the worrying realization that, sometimes, architects and, more in general, professionals in the field of building technology overlook some important aspects, passing very late in the design process the responsibility of building envelope performance, often demanding it to mechanical and building envelope engineers.

Frequently, the performance response expected during the design phase does not match with what was obtained during the construction phases. Detailed design decisions are essential from the very beginning of the process to ensure the correspondence among initial concept and its realization, especially when dealing with advanced technologies for building envelope or with specific building components that need particular attention during the design and installation phase, such as—for instance—transparent building components (Fig. 9.1).

Understanding their nature and behavior within the whole system allows conceiving a new generation of architectural solutions able to provide an operational response balanced among performance to be guaranteed and regulations to be respected, here defined with the term "Smart Architecture."

So, first of all, the present work aims at defining one of the multiple facets of the concept of "smart" in Architecture, for which many interpretations are existing, but a precise definition of its meaning appears surprisingly difficult. In fact, even if today, the smart feature in the architecture's domain is generally associated with materials and technologies highly engineered; this concept has not been intended only as the integration of different or unconventional features in a given material, component, or system, but it can also be attributed to the intelligent application of something within building system in its whole or, again, to the intelligent management of a complex organism, such as the building one.

Accordingly, such an idea led to the definition of "Smart Buildings" as systems able to adapt—in a responsive way—not only to the variability of boundary conditions but also to the continuously evolving needs of their end-users.



V. Frighi, *Smart Architecture – A Sustainable Approach for Transparent Building Components Design*, https://doi.org/10.1007/978-3-030-77606-0_9

**Fig. 9.1** Curious resolution of the "interface criticalities" in a traditional existing window in Italy. Photo credit: the author

Thus, a Smart Building is not only an edifice that makes use of Building Automation Technologies—definition which sounds quite reductive—but rather it is a building equipped with technological solutions able to provide high performance in terms of comfort, energy efficiency, and environmental sustainability, capable of dynamically interacting with the surrounding and to acquire data and information useful for optimizing its functioning, thus correct possible inefficiencies.

In this framework, the development of a supporting database can help in providing a systematic characterization of Smart Architecture for future trends in design, mapping strategies adopted in such systems, recognizing common patterns, and identifying unexplored concepts to help further development of high-potential, innovative smart building solutions.

Besides, this recognition has been necessary to acquire knowledge about the functioning of Smart Buildings and to understand how the advancements in technologies here analyzed have changed architectural features over time becoming, in some cases, innovations of specific traditional elements. Indeed, the establishment of architectures with responsive and adaptive capabilities, previously unthinkable, to face the modern need of change, has rapidly triggered a revolution in the creation of the architectural work as a whole, giving technology for architecture an active role: no longer as "engineering" of a project but rather as an element of synthesis—centralizer of multiple instances—on which leverage to foster interactions among buildings, surrounding environment, and user needs.

It has been evident that the application of the innovations here investigated has produced a modification of architectural practices, enriching design language through the adoption of heterogeneous components, sometimes known but endowed with new technological values, while sometimes resulting from technological transfer processes from other sectors or the most recent scientific and technical discoveries.

The evolution of the demanding framework that modified the use of materials and products according to a functional approach toward matter (Sinopoli and Tatano 2012) generated architectures with complex shapes able to overcome the simplification of traditional architectural solutions, often determined by the assembly of regular elements, in favor of more evolved shapes, assisted by ICT for the building process, thus making building materials functional elements that could perform an adaptive behavior.

So, Smart Architecture, whose technological components acquire new aesthetic-formal value, becomes the driving force for the development of unconventional ideas and forms, being "the next big milestone" of the technology for architecture (Gallo and Romano 2016), thanks to their ability in giving buildings the desired resilience capacity becoming, at the same time, iconographic supports able to evolve with the use they made of innovative materials and components (Romano 2013).

Therefore, even the use of materials has been separated from their application, according to conventional forms and meanings dictated by cultural traditions, availability of resources or precise formal and aesthetic features, to become linguistic means for communicating the project, with precise functional intentions, expression of a new *glocal* (global + local) language, able to adapt to the ever-changing needs of a hybrid contemporary society, while finding at the same time a relationship with local traditions, even if innovated in form and vocation.

Concerning the choice to further focus the work on transparent building components, it has been made due to fenestration design issues, which still constitute a gap in the broad scenario of architectural design due to their significant role in building energy efficiency and comfort of the internal environment.

Although nowadays the market offers different types of high-performance glazing systems, ever more frequently enhanced with added functions to increase their whole performance response as building envelope components, Smart Windows and, more in general, windows technologies with outside-the-box performance are still in the early stage of development.

So far, transparent building components—even if equipped with quite high-performance—are still the weakest point of building envelopes, especially in terms of thermal resistance.

For this reason, fenestrations' design optimization challenge is more complex than for many other building components and systems, so smart windows technologies are still often neglected in architecture and buildings, although some functions would already be technically feasible.

Research is now focusing on strategies to improve windows' performance, reducing costs, and favoring sensors and control systems installation to increase their operation efficiency, making them monitorable, and reactive to various stimuli

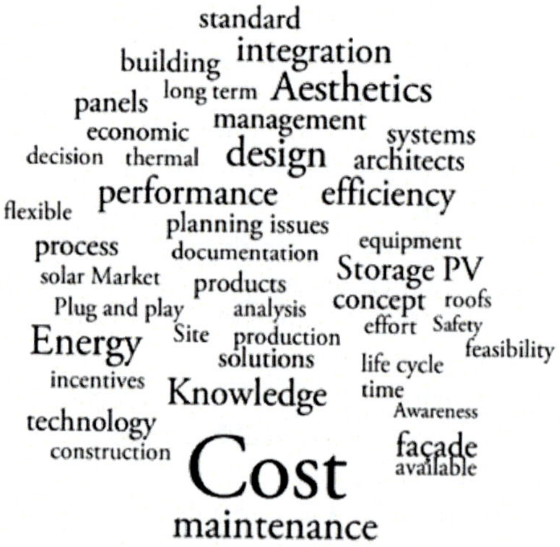


**Fig. 9.2** Keywords density of barriers declared by experts for widespread integration in building envelope of smart windows technologies. Author reworked version on Prieto et al. (2017)

based on predetermined inputs, quantifying at the same time benefits thus obtained. Developing windows with reduced weight, cost, and thermal features, able to control and harvest energy, is of paramount importance because they will have a high impact on the window industry (Fig. 9.2).

However, at the same time, there is the need to provide designers with appropriate tools to manage such complexity, according to a holistic and integrated sustainable approach that considers building systems as their whole.

Therefore, after collecting and presenting the state of the art about available smart windows' technologies, this work results in the development of a tool(box) to support decision-making toward the conscious and effective integration of transparent building components within smart building envelope systems. It gathered results and considerations coming from preliminary literature and technical reviews to further re-elaborate and systematize them, helping designers and professionals in the fields in developing new high-performance solutions, thus solving the problem of design and construct sustainable smart buildings.

The tool(box) is composed of a *balanced scorecard*, which helps in defining the set of actions to be undertaken to increase performance features of products evaluated, about predetermined requirements; an *assessment matrix*, which systematizes measures depicted within the balanced scorecard with performance objectives; and a *Smart Windows' Configurator*, an open matrix for compiling and validating different technical solutions, supporting the design of transparent building components toward the improvement of building envelope performance.

By using elements coming from the previous analysis, such a tool focuses on suggesting a variety of near-optimal possibilities based on the optimization of single building envelope requirements. It has been proved that suggesting multiple and diverse good initial alternatives is more beneficial for the early design stages when designers are still exploring various design routes and numeric detailed properties of any building elements systems are irrelevant.

The general approach used by the tool foresees the introduction of some input data, related to the climatic condition of the application surroundings, to the type of building under investigation, and to further select the type of windows technologies for which advice is needed. Interpreting such data, already known at the initial design stages, general design considerations related to building requirements implementation are provided.

For synthesis purposes, in the present book, final concepts are stated only for the combinations among input data recognized as the most frequent by the sample analysis performed within Chap. 4, pointing the attention toward technologies that obtained the lower score within the assessment matrix depicted in Chap. 8, Sect. 8.1. The structure of the configurator is suitable to further implementation according to the various possible combinations of the variables involved. At the moment indeed, it is designed as a standalone tool, suitable to be translated into software for an easier consultation, able to provides design directions and assist architects in the early design stages of a smart building envelope or, again, to support researchers and producers in deciding on the most pertinent research and development activities concerning windows technologies.

This work wants to show that a smart building envelope can only be the result of an informed design process that requires very careful attention since its very first stages. Yet, some differences distinguish these actions from the "traditional" design activity. Smart architectural design decisions, at the whole building level, must be considered and passive design strategies aimed at achieving and ensuring the occupants' comfort; the choice of the most suitable elements coupled with their proper integration within the main building envelope interfaces allow to make buildings adaptable to the environmental challenges we are facing.

Moreover, it is important to clarify that not all the building elements and materials chose must have an outstanding performance—as it sometimes adds unwanted complexity to the strategies put in place—but rather they must be based on conscious and sustainable design decisions, being considered in their whole and in relationship with the other components installed. It has to be considered that even the buildings selected as examples of Smart Architecture cannot be obtained resorting only to a collection of "smart" elements placed together, then somehow expected to perform adequately; also, components with an extremely high degree of smartness require a proper integration, programming, and control to face the different challenges they are going to meet.

This makes it essential to adopt strategic addresses for building applications of the wide variety of materials and components available for smart building envelopes such as advanced integrated facades, adaptive glazing, thermal massive components, dynamic insulation systems, and so on. Only their strategic and well-studied combination can enrich the design, providing a real implementation of building envelope performance in their whole.

Furthermore, nowadays, the design of a smart building envelope can benefit from modern design tools that have taken advantage of the computer development of the fourth industrial revolution. This has expanded the range of available possibilities, informing new design options and allowing to perform corrective operation also before construction, even if it has to be said that software and digital tools currently available generally evaluate finished alternatives rather than suggest design directions.

On the contrary, the tool(box) developed within the present work aimed at encouraging the arising of relevant questions ("what if" scenarios) and suggests a series of solutions or possible promising design directions according to the specific needs that concern the design of transparent building components. It will also trigger the further development of advanced tools that can enable the exploration of different novel solutions through a smart approach.

Indeed, it can be expected that smart building envelopes will have a much wider implementation in the near future due to the acceptance—within the construction industry—of technological innovations of last years. Any forthcoming directions for the development of high-performance smart building envelopes must include an improved cognition of perception elements, as well as the building industry, need to handle more detailed information beyond what is available today.

Certainly, further research has to be done so that smart building components can be expected to become equipped with devices based on wireless data transmission—such as, for instance, intelligent sensors capable of detecting real-time data about environmental parameters, consumption, user behavior, and operation of technical plants—which allow to concretely suppose the development of a new generation of products, even smarter than before, capable of autonomous and efficient management. Additionally, obtaining feedbacks and real-time information on them will lead to more informed decision makers and to the definition of more effective management strategies that will be reflected on a real implementation of the systems thus obtained, making them real User-Generated Contents (UGCs), thus guaranteeing their functionality over time, and, above all, their compliance with the ever-changing needs of their occupants, making them (inter)active and interoperable structures, thanks to the introduction of the aforementioned technological innovations, as an element of mediation among multiple instances, sometimes juxtaposed but essentially interconnected.

Further studies are then required to provide new operation modes to well-known actuators, such as blinds, shading elements, lighting controls, and so on. Other directions for research include new additional explorations in the fields of smart materials, accomplishing a series of functions that currently cannot be carried efficiently with traditional materials and components.

Moreover, ongoing challenges are those related to the proper design of building envelope interfaces, especially when dealing with such advanced technologies, for their nature equipped with high performance, to guaranteeing their performance maintenance over time. Indeed, detailed design of heterogeneous components as a whole constitutes the contact points among the conceptual design phase and the implementation, actualizing one, means to avoid the performance decay over time.

## References

Gallo P, Romano R (2016) Adaptive facades, developed with innovative nanomaterials, for a sustainable architecture in the Mediterranean area. Paper presented at the International High-Performance Built Environment Conference, Sydney, 2016

Prieto A, Knaack U, Auer T, Klein T (2017), Solar façades – Main barriers for widespread integration of solar technologies, Journal of Façade Design & Engineering, Vol. 5, N. 1

Romano R (2013) Sistemi di chiusura a comportamento dinamico. In Scudo G (2013) (ed), Tecnologie solari integrate nell'architettura. Processi, Strumenti, Sistemi, Componenti. Wolters Kluwer, Milanofiori Assago, pp. 181-193

Sinopoli N, Tatano V (2012) Sulle tracce dell'innovazione. Tra tecniche e architettura. FrancoAngeli, Milano

# Index




V. Frighi, *Smart Architecture – A Sustainable Approach for Transparent Building Components Design*, https://doi.org/10.1007/978-3-030-77606-0